T0262120

Soil Fertility

Soil Fertility

Edited by **Lester Bane**

New York

Published by Callisto Reference,
106 Park Avenue, Suite 200,
New York, NY 10016, USA
www.callistoreference.com

Soil Fertility
Edited by Lester Bane

© 2015 Callisto Reference

International Standard Book Number: 978-1-63239-563-4 (Hardback)

This book contains information obtained from authentic and highly regarded sources. Copyright for all individual chapters remain with the respective authors as indicated. A wide variety of references are listed. Permission and sources are indicated; for detailed attributions, please refer to the permissions page. Reasonable efforts have been made to publish reliable data and information, but the authors, editors and publisher cannot assume any responsibility for the validity of all materials or the consequences of their use.

The publisher's policy is to use permanent paper from mills that operate a sustainable forestry policy. Furthermore, the publisher ensures that the text paper and cover boards used have met acceptable environmental accreditation standards.

Trademark Notice: Registered trademark of products or corporate names are used only for explanation and identification without intent to infringe.

Printed in the United States of America.

Contents

Preface

Every book is a source of knowledge and this one is no exception. The idea that led to the conceptualization of this book was the fact that the world is advancing rapidly; which makes it crucial to document the progress in every field. I am aware that a lot of data is already available, yet, there is a lot more to learn. Hence, I accepted the responsibility of editing this book and contributing my knowledge to the community.

A descriptive account based on the extensive topic of soil fertility has been presented in this book. It is a compilation of topics presented by internationally acclaimed experts in the field of soil fertility. It discusses in detail about various biological processes along with the utilization of organic and inorganic fertilizers and how they contribute towards improving soil fertility. It also includes discussions on improving fertilizer recommendation and efficiency and elaborates leaf sampling and analysis. Standardized methods of analysis and proper leaf sampling play important roles in providing good recommendations.

While editing this book, I had multiple visions for it. Then I finally narrowed down to make every chapter a sole standing text explaining a particular topic, so that they can be used independently. However, the umbrella subject sinews them into a common theme. This makes the book a unique platform of knowledge.

I would like to give the major credit of this book to the experts from every corner of the world, who took the time to share their expertise with us. Also, I owe the completion of this book to the never-ending support of my family, who supported me throughout the project.

Editor

Biological Processes and Integration of Inorganic and Organic Fertilizers for Soil Fertility Improvement

Improvement of Soil Fertility with Use of Indigenous Resources in Lowland Rice Systems

Satoshi Nakamura, Roland Nuhu Issaka,
Israel K. Dzomeku, Monrawee Fukuda,
Mohammed Moro Buri, Vincent Kodjo Avornyo,
Eric Owusu Adjei, Joseph A. Awuni and
Satoshi Tobita

Additional information is available at the end of the chapter

1. Introduction

Low inherent soil fertility has been identified as a major cause for low rice yield in Sub-Saharan Africa (Buri et al., 2004; Senayah, et al., 2008; Buri et al., 2009; Issaka et al., 2009; Abe et al., 2010). The problem is compounded because farmers are not able to purchase fertilizer due to relatively high cost and therefore rely mostly on natural soil fertility which is low and declining. However, there are various organic materials that have the potential and can effectively contribute to improving soil fertility within the region. JIRCAS (2010) reported the quantity, quality and distribution of various organic materials that are available and suitable for supplementing soil fertility in Ghana. In this report, Rice straw (RS), Cow dung (CD), and Human excreta (HE) were evaluated in the Northern region which is located within the Guinea savanna zone and found them effective materials that can contribute in reducing chemical fertilizer application, and to improve soil fertility. RS, Poultry manure (PM), and Saw dust (SD) were evaluated in the Ashanti region located within the Equatorial forest zone. These two climate zones are common agricultural zones where lowland rice is cultivated in Africa.

RS, the commonest material within the two regions, is one of the most accessible materials to resource poor peasant farmers, because RS is produced in rice fields itself and therefore does not need to be transported. Therefore, the development of proper and improved management techniques of this material is essential for Ghanaian rice production.

In the Northern region, CD and HE were selected as regional materials for agricultural use. These materials have been evaluated and used in some Asian countries as having high fertilizer effects (e.g. Austin et al., 2005, Matsui, 1997). However, in Ghana, these materials are considered as non-accessible resource because of lack of proper management technology, and psychological reasons (Coffie et al., 2005). In Ghana, it seems that gathering plant materials in the Northern region is more difficult than in the Ashanti region. Human and/or animal resource could be collected in settlement ecosystem through some life style or system innovations.

In the Ashanti region, PM and SD were selected. These materials are considered as having high agronomical potential especially in the Ashanti region. Commercial poultry farmers are generally concentrated in the Greater Accra and Ashanti regions hence almost 50% of poultry manure is produced in these two regions (Quarcoo, 1996). So poultry manure is particularly accessible and available for rice production in the Ashanti region. SD is a byproduct material of lumbering industry which is popular in Ghana. Hence, its disposal has lately become a major problem for the government and timber industry.

This study therefore investigated the application effect of these selected organic materials on rice yield, and their proper processing methods for each organic matter application. The various processing (referred here as pre-treatment) include the following four treatments i.e. (i) ashing, (ii) charring, (iii) composting, and (iv) direct application. The effect of these four treatments on rice growth and yield were investigated. Organic materials have various forms, and each form has some advantages and disadvantages, as far as crop nutrition is concern. Therefore, this study investigated the best organic matter management option for rice production in Ghana, through the comparison of rice yield under four types pretreated i.e. ashed (ASH), charred (CH), composted (CO), or non-treated raw material (RW) organic materials application.

2. Materials and methods

2.1. Research site

Ghana has several agro-ecological zones but these can be broadly categorized into forest and savannah. The experiments were conducted at the University of Development Studies (UDS; N 09°24'19", W 000°58'14") located in the Northern region which lies within the Guinea savannah agro-ecological zone, and at the CSIR - Soil Research Institute (SRI; N 06°45'18", W 001°35'30") which is in the Ashanti region and within the forest agro-ecological zone of the country. At UDS, the effect of rice straw (RS), cow dung (CD), and human excreta (HE) application on rice yield were investigated. At the CSIR-SRI, the effect of RS, poultry manure (PM), and saw dust (SD) application on rice production was evaluated.

2.2. Treatments

Three types of organic materials that are potentially available were selected, for study site. Each of the organic material was applied to rice on the fields after pretreatment into ash (ASH), charr (CH), compost (CO), and untreated raw material (RW). The effect of Phosphate

Rock (PR) on rice yield was also investigated by setting-up three levels of phosphorus i.e. CON as control (0 kgP$_2$O$_5$ ha^{-1}), BRP as (Burkina Faso phosphate rock) applied at 135kg P$_2$O$_5$ ha^{-1}, and TSP as mineral fertilizer (135kg P$_2$O$_5$ ha^{-1}). 36 treatments (3 x 4 x 3) and without-organic material (WOM) application plot at 3 levels of P application and three replication., This gave a total of 117 plots in each station.

The quantities of the three types of organic materials applied were maintained as 3.2 kg P$_2$O$_5$ ha^{-1} at the Northern site, and applied directly to the soil surface. Compost was purchased from the market because there was not enough compost at that time. In Ashanti region, 3.2 kg P$_2$O$_5$ ha^{-1} of organic materials were applied as direct application, and 1 t ha^{-1} of materials were applied for pretreated organic materials. For all treatments, 30 kg N ha^{-1} and 30 kg K$_2$O ha^{-1} was applied as basal dressing.

Rice varieties in our trials were selected GR18 in Northern region, and Sikamo in Ashanti region. GR18 is reported to be one of the most popular varieties in Northern Ghana (Ghana Seed Company, 1988). Sikamo was recommended by Crops Research Institute (CRI), and was an improved variety for rain-fed lowlands rice cultivation, demonstrated in the 1990s by Ghana rice project.

The plant density was at a recommended rate of 20 cm × 20cm in each site.

3. Results

A 3-way analysis of variance (ANOVA) for rice grain yield under 36 treatments was conducted to verify the effect of organic matter application at each site, and to clarify the effect of interaction between three factors, i.e. type of organic material, pretreatment, and PR application. In this analysis, rice yield without organic matter application was excluded. The results of the 3-way ANOVA for rice grain yields did not show any significant difference between treatments and their interaction (Tables 1 and 2). Even though, all treatments were replicated three times, but it might not be enough to avoid specific variance of land condition for elucidation of statistical difference.

Factor	Degrees of freedom	Sum of squares	Mean square	F value	lsd(5%)	F(5%)	significance
A: Type of organic material	2	0.41	0.21	0.07	0.79	3.13	ns
B: Pretreatment	3	1.31	0.44	0.16	0.91	2.74	ns
C: Phosphate rock application	2	0.61	0.30	0.11	0.79	3.13	ns
A×B	6	5.04	0.84	0.30	1.57	2.23	ns
A×C	4	0.83	0.21	0.07	1.36	2.50	ns
B×C	6	2.38	0.40	0.14	1.57	2.23	ns
A×B×C	12	10.45	0.87	0.31	2.73	1.89	ns
Error	70	196.10	2.80				

Table 1. 3-way ANOVA for rice grain yield in the Northern region

Factor	Degrees of freedom	Sum of squares	Mean square	F value	lsd(5%)	F(5%)	significance
A: Type of organic material	2	3.44	1.72	0.08	2.12	3.13	ns
B: Pretreatment	3	3.84	1.28	0.06	2.45	2.74	ns
C: Phosphate rock application	2	11.86	5.93	0.29	2.12	3.13	ns
A×B	6	3.76	0.63	0.03	4.24	2.23	ns
A×C	4	0.46	0.11	0.01	3.67	2.50	ns
B×C	6	7.50	1.25	0.06	4.24	2.23	ns
A×B×C	12	12.21	1.02	0.05	7.34	1.89	ns
Error	70	1421.13	20.30				

Table 2. 3-way ANOVA for rice grain yield in the Ashanti region

3.1. Effect of indigenous organic resources application on rice yield in Northern region

Rice yield under various organic material applications are indicated in Table 3. Least significant differences (LSD) at 5 % level are also shown in the Table. Statistical difference analysis, however, accord ANOVA priority to LSD. Mean value of rice grain yields under combined application of organic material, pretreatment, are indicated in Figure 1. The mean values were calculated by averaging of CON, BRP, and TSP for the three replications. The rice yield in Northern region generally showed higher value (e.g. 2.51 t ha^{-1} at CO-HE mean value of CON, BRP, and TSP) compared with WOM (1.25 t ha^{-1} in means value of CON, BRP, and TSP).

CO and RW showed relatively higher value in RS treatments. It seemed that ASH and CH treatment for rice straw had lost more nutrients than the other organic materials. Meanwhile, RW-RS and CO-RS have physical beneficial effect on soil fertility such as mulching effect.

In the CD treatments, ASH and CH showed higher value in CON treatment, and CO and RW showed higher value in BRP treatment. CD contained highest P_2O_5 contents (1.72%) among three selected organic material. It maybe considered as the contribution of mineralized P from CD under combustion, that contributed to the increased rice grain yields.

The rice yields under HE application showed high value in CH and CO pretreatments. Especially under CON treatment, CH showed 2.97 t ha^{-1} of grain yield and CO showed 2.93 t ha^{-1} of grain yield, respectively.

Generally, organic materials application increased rice grain yields compared with WOM. Organic materials application showed pronounced effect especially in CON treatment, i.e. in RS 3.96 times, in CD 4.21 times, in HE 4.46 times higher than WOM rice yield, respectively. On the other hand, there were not significant effects of organic material applications on rice yield under TSP treatment. The rice yield under TSP treatment ranged from 82 to 91% against rice yield on WOM.

Ashing has disadvantageous of nutrient loss and/or green-house gas production with dry combustion. Meanwhile when the material is ashed, there is significant reduction in both weight and volume. Labour and cost of transportation is greatly reduced resulting in economic benefit and transportation efficiency. Ash contributes in the enhancement of soil fertility with mineralization of organic matter, in elimination of bad organic substances that may become plant growth inhibition factor, in supplement of Potassium and Silica (Anzai 1993).

Charring has similar advantages and disadvantages as in ashing in the short-term but is expected to show organic function compared with ashing. Recent studies reported that application of charred organic material could enhance crop production through increasing of soil organic matter, improvement of soil physical and biological properties (Glaser et al., 2002; Lehmann et al., 2003; Yeboah et al., 2009). And recently, the application of charred organic material has been focused as one of the important factor on the pedogenic process of Amazonian Dark Earth, that is known as tropical fertile soil with black colored surface (Nakamura et al., 2007)

It is well known that composted organic materials strongly contribute to crop production. Composting enhances the fertilizing capacity of an organic material. The C/N ratio of an organic material decreases with composting, mineralization of the material also results in mineral nutrient concentration. Organic matter decomposition and fermentation, can reduce disease and pestilence risk due to increase in temperature. However, it is also known that composting requires knowledge for keeping fermentation condition i.e. water contents management. Composting is also laborious and the material is normally bulky requiring additional labour to convey the material to the field and apply. Therefore it may be a difficult technical option for farmers due to some of these reasons.

	Rice Straw (RS) management				Cow Dung (CD) management				Human Excreta (HE) management			
	CON	BRP	TSP	Ave	CON	BRP	TSP	Ave	CON	BRP	TSP	Ave
WOM	0.50	1.14	2.09	1.25	0.50	1.14	2.09	1.25	0.50	1.14	2.09	1.25
ASH	1.53	1.99	1.65	1.72	2.48	1.83	1.92	2.08	1.80	1.13	1.83	1.59
CH	1.46	1.51	1.87	1.61	2.62	1.88	1.63	2.04	2.97	1.76	1.24	1.99
CO	2.40	1.92	1.74	2.02	1.96	2.32	1.67	1.98	2.93	2.54	2.06	2.51
RW	2.61	1.57	2.38	2.19	1.43	2.68	1.67	1.93	1.31	1.82	2.34	1.83
Ave	1.70	1.62	1.95	1.76	1.80	1.97	1.80	1.86	1.90	1.68	1.91	1.83
LSD(5%)	1.70	1.42	1.67	0.86	1.90	2.27	1.29	0.96	1.97	0.92	1.62	0.88
Effectiveness of organic matter application												
WOM	1.00	1.00	1.00	1.00	1.00	1.00	1.00	1.00	1.00	1.00	1.00	1.00
ASH	3.03	1.74	0.79	1.38	4.92	1.61	0.92	1.67	3.57	0.99	0.87	1.27
CH	2.89	1.32	0.89	1.29	5.20	1.65	0.78	1.64	5.88	1.55	0.59	1.60
CO	4.76	1.69	0.83	1.62	3.89	2.04	0.80	1.59	5.80	2.23	0.98	2.01
RW	5.18	1.38	1.14	1.76	2.84	2.35	0.80	1.55	2.60	1.60	1.12	1.47
Ave	3.96	1.53	0.91	1.51	4.21	1.91	0.82	1.61	4.46	1.59	0.89	1.59

Grain yield: (t ha^{-1}); CON: Non-P application, BRP: Burkina Faso Phosphate Rock application, TSP: Triple Super Phosphate application, WOM: without organic material, ASH: ashing, CH: charring, CO: composting, RW: raw material,

Effectiveness of organic matter application: The rice grain yield at each plot / the rice grain yield at WOM under same P management condition.

Table 3. Rice grain yield under organic and pretreated materials in Northern region

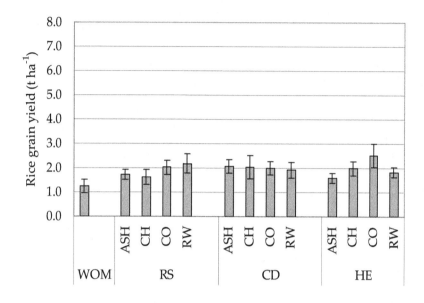

Figure 1. Effect of organic matter application on rice grain yield in Northern region. Mean values of CON, BRP, and TSP treatment are indicated. WOM: without organic matter, RS: rice straw, CD: cow dung, HE: human excreta, ASH: ashing, CH: charring, CO: composting, RW: raw material

3.2. Effect of indigenous organic resources application on rice yield in Ashanti region

Rice yields under various organic materials applications in Ashanti region are shown in Table 4 and Figure 2. The significant difference by 3-way ANOVA was not found as the same as Northern region. The cultivar used in Ashanti region (Sikamo) was different from Northern region (GR18), so rice yields could not be simply compared between the two research sites.

In the RS treatment, the mean rice yield values for CON, BRP, TSP rice yields showed higher value than WOM, but RW-RS showed relatively lower values than the other treatments. The RW-RS was the best harvested RS treatment in Northern region, so there were different effectiveness trend of RS treatment between Northern and Ashanti region. Northern region is located in Guinea-Savannah zone, and has little annual precipitation. Soils were kept under upland condition over a long time. Ashanti region is in the Equatorial Forest zone that has higher rainfall and that lowland rice can be cultivated under submerged condition throughout the year. Probably the difference of RW-RS effect on rice yield between the two sites was attributed to difference of organic matter decomposition rate caused by variation in water conditions.

	Rice Straw (RS) management				Poultry Manure (PM) management				Saw Dust (SD) management			
	CON	BRP	TSP	Ave	CON	BRP	TSP	Ave	CON	BRP	TSP	Ave
WOM	5.10	6.20	5.27	5.52	5.10	6.20	5.27	5.52	5.10	6.20	5.27	5.52
ASH	6.03	6.43	7.27	6.58	5.20	6.83	7.20	6.41	5.63	6.80	6.47	6.30
CH	6.67	6.10	7.27	6.68	5.93	6.17	7.27	6.46	6.17	6.63	6.13	6.31
CO	5.73	7.27	6.93	6.64	6.43	7.13	6.70	6.76	6.03	5.57	7.87	6.49
RW	6.30	6.47	6.20	6.32	6.20	6.73	6.77	6.57	5.33	5.90	5.17	5.47
Ave	5.97	6.49	6.59	6.35	5.77	6.61	6.64	6.34	5.65	6.22	6.18	6.02
LSD(5%)	0.95	2.14	2.05	0.95	1.03	1.80	1.93	0.92	1.23	2.18	1.78	1.04
Effectiveness of organic matter application												
WOM	1.00	1.00	1.00	1.00	1.00	1.00	1.00	1.00	1.00	1.00	1.00	1.00
ASH	1.18	1.04	1.38	1.19	1.02	1.10	1.37	1.16	1.10	1.10	1.23	1.14
CH	1.31	0.98	1.38	1.21	1.16	0.99	1.38	1.17	1.21	1.07	1.16	1.14
CO	1.12	1.17	1.32	1.20	1.26	1.15	1.27	1.22	1.18	0.90	1.49	1.18
RW	1.24	1.04	1.18	1.14	1.22	1.09	1.28	1.19	1.05	0.95	0.98	0.99
Ave	1.21	1.06	1.31	1.19	1.17	1.08	1.33	1.19	1.14	1.00	1.22	1.11

Grain yield: (t ha⁻¹); CON: Non-P application, BRP: Burkina Faso Phosphate Rock application, TSP: Triple Super Phosphate application, WOM: without organic material, ASH: ashing, CH: charring, CO: composting, RW: raw material,

Effectiveness of organic matter application: The rice grain yield at each plot / the rice grain yield at WOM under same P management condition.

Table 4. Rice grain yield under organic and pretreated materials in Ashanti region

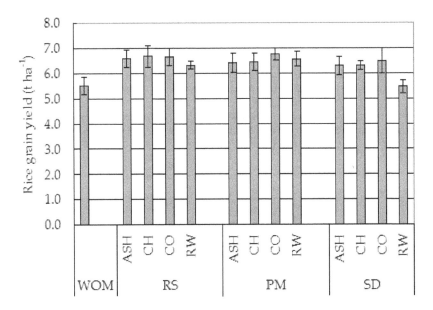

Figure 2. Effect of organic matter application on rice grain yield in Ashanti region. Means values of CON, BRP, and TSP treatment are indicated. WOM: without organic matter, RS: rice straw, PM: poultry manure, SD: saw dust ASH: ashing, CH: charring, CO: composting, RW: raw material

The application of poultry manure resulted in higher yield except ASH and CH under CON treatment. However, ASH-PM showed low yield (5.20 t ha^{-1}) under CON treatment, but showed high yield under BPR and TSP treatments (6.83 and 7.20 t ha^{-1}), respectively.

Under the SD application, RW-SD produced a lower value compared with other SD application. RW-SD under all three P treatments showed a similar yield as WOM (Effectiveness against WOM: 1.05 at CON, 0.95 at BRP, 0.98 at TSP). Only the RW-SD did not show the positive effect of organic matter application on rice yield. The RW-SD in Ashanti region was direct application of woody material for submerged condition. So, decomposition rate and plant nutrient release rate seemed to be extremely slow, and thus causing nitrogen starvation for the treatment.

3.3. The combination effect of various organic materials and phosphate rock application

The agronomic efficiencies to TSP under various organic matter applications are shown in Table 5. Agronomic efficiency was calculated from the difference in grain yield (t ha^{-1}) of PR to those of TSP treatment. This efficiency can be considered as the indicator of BRP being a possible alternative to TSP. According to FAO (2004), BPR direct application showed 97% of agronomic efficiency in lowland rice cultivation.

Under this study and within the Northern region in Ghana, the agronomic efficiency of BPR direct application against TSP was relatively lower in WOM, RW-RS, ASH-HE (54, 66, 62%, respectively). Treatments of ASH-RS, RW-CD, CO-CD, CH-HE, CO-HE showed efficiency of over 120%, suggested that combined application of PR and organic materials was effective in PR application. It is well known that organic matter application is one of the effective techniques to solubilize PR. This may be so due to PR dissolution in organic acids produced by microorganisms, through decomposition of easily decomposable organic substances.

On the other hand, within the Ashanti region, only CO-SD showed a lower value (71%). The other treatments including WOM gave similar trend with higher values that ranged from 84% to 118%. This result shows that agronomic efficiency did not increased by organic matter application. These results do not only indicated that BPR can be use directly as alternative fertilizer, but also indicated that there are not clear positive effects of organic matter application on PR agronomic efficiency enhancement.

	Northern			Ashanti		
	RS	CD	HE	RS	PM	SD
ASH	121	95	62	89	95	105
CH	81	115	142	84	85	108
CO	111	139	123	105	106	71
RW	66	160	78	104	100	114
WOM	54			118		

RS: rice straw, CD: cow dung, HE: human excreta, PM: poultry manure, SD: saw dust, WOM: without organic material, ASH: ashing, CH: charring, CO: composting, RW: raw material directly applied

Table 5. Effect of various organic material application on changes of agronomic efficiency (%) in BPR direct application against TSP application

4. Conclusions and recommendations

Rice grain yields generally increased with the application of the various organic matters. Grain yields under organic matter application showed approximately 1.5 to 1.6 times higher than those of WOM in Northern region, and 1.1 to 1.2 times higher than WOM in Ashanti region. However in some case e.g. RW-SD, did not show any positive effect of application on rice yield, and so the proper pretreatment for each organic resources need to be selected.

4.1. Suggestion for indigenous organic material application in Northern region, for rain-fed lowland rice cultivation in Savannah zone

The best promising organic material for agricultural use in Northern region is rice straw (RS). CO-RS and RW-RS treatment showed high positive effect on rice yield in the Northern region trial. It seems adequate amount of mineral plant nutrients were supplied due to decomposition of RS and that soil physical and biological properties were enhanced. On the other hand, CH-RS and ASH-RS also showed positive effect on rice yield, however, it should be considered that charring needs input of labor and capital, and that ashing will impact the environmental negatively. Moreover, ashing process can reduce effectiveness of soil physical and biological properties observed under CO and RW application.

CD application showed high yield in all pretreated plots. It is well known that CD application is effective for rice production. In the Northern region, however, it is far from recommended materials because of difficulty in material gathering. Most of CD in this region was produced by cattle grazing, that means farmer needs to collect CD scattered throughout savannah forest. This study indicated that CD has similar effect on rice yield compared with other organic materials. The use of CD in rice cultivation can be as popular as RS due to both availability and accessibility. While RS is readily available in the rice fields extra labour is require to search and collect CD. However, after using CD as fuel CH-CD and ASH-CD can be used in rice production since these materials increased rice yield significantly.

Rice yields average of HE application was almost same as those of CD. Improvement of crop productivity has been reported by many existing studies. However it also has been mentioned that HE usage has difficulty to diffuse for farmers because of hygienic risk and psychological avoidance. These obstacles can be resolved by either combustion process such as CH and ASH, or composting process. CO-HE indicated highest yield among four pretreatment in HE treatments, the yield was almost twice higher than WOM.

According to the results from Northern region trial, authors suggest that CD and/or HE composting based on RS usage should be well examined as the effective and affordable technical options for farmers. HE composting still have the difficulty of gathering, but technology introduction on collecting and separating urine and feces, such as Eco-San toilet, will open the way for proper management of HE, especially in Northern region. Moreover CD and HE composting based RS can be expected to increase RP application effect on rice yield.

4.2. Suggestion for indigenous organic material application in Ashanti region, under water controlled lowland rice cultivation in Equatorial forest zone

Similar to what was observed in Northern region, the use of RS is considered as an effective resource for rice production in Ashanti region, because of its high accessibility. However, unlike the Northern region, RW-RS in Ashanti region showed a smaller effect on rice production than the other treatments. Most rice fields in Ashanti region are irrigated and can be used to cultivate throughout the year. Therefore, rice fields are generally maintained under submerged condition for long periods hence decomposition rate of organic matter will be relatively slow. Hence, RW-RS application is discouraged because its effect on rice yield is not pronounced, and a possible risk of nitrogen starvation.

Poultry manure with every pretreatment showed high positive effect on rice yield enhancement, especially in CO-PM. Poultry manure is a popular organic material that is an effective P source, and that is highly accessible in the Ashanti region.

In the SD treatment, RW-SD showed lower value (5.47 t ha^{-1}) than WOM of rice yield (5.52 t ha^{-1}). Similar to RW-RS, direct application of SD generally has the problem due to its high C/N ratio. Moreover SD composting is also difficult because SD consist of woody material that contains resistant organic matter. To avoid plant damage by organic acids and/or nitrogen starvation, SD composting needs to take a long time for decomposition and fermentation. Combustion treatment, which take shorter time and is easy to practice (CH and ASH) is affordable and effective for agricultural usage in Ashanti region. The possibility of composting through inoculation with microorganisms should not be excluded from future investigation.

In the organic material application in Ashanti region, all treatment except direct application of RS and SD showed positive effect on rice grain yield. It is suggested that ASH-RS, CH-RS, ASH-SD, CH-SD, and composting RS or SD combined with PM, and also RW-PM, are evaluated as promising technical options that are accessible and effective for rice cultivation in Ashanti region. These selected options showed relatively high value not only in rice yield but also in agronomic efficiency of BPR application. The effect of organic material application on enhancement of PR solubility was smaller compared with the effect observed in Northern region. However, direct application of PR in this region indicated high agronomical efficiency to TSP application, and means PR has possibility of alternative usage against TSP. Further investigation need to be conducted on the effect of RP combined with organic material application on rice yield.

Acknowledgement

This investigation was conducted under the project "Improvement of Soil Fertility with Use of Indigenous Resources in Rice Systems of Sub-Sahara Africa', which was funded by the Ministry of Agriculture, Forestry and Fisheries (MAFF), Japan. The authors are grateful to the entire staff of CSIR-SRI, UDS both in Ghana and JIRCAS for their technical supports.

Author details

Satoshi Nakamura[1], Roland Nuhu Issaka[2], Israel K. Dzomeku[3], Monrawee Fukuda[1], Mohammed Moro Buri[2], Vincent Kodjo Avornyo[3], Eric Owusu Adjei[2], Joseph A. Awuni[3] and Satoshi Tobita[1]

1 Japan International Research Center for Agricultural Sciences (JIRCAS), Ohwashi, Tsukuba, Ibaraki, Japan

2 CSIR-Soil Research Institute (SRI), Academy Post Office, Kwadaso - Kumasi, Ghana

3 University for Development Studies (UDS) Faculty of Agriculture, Department of Agronomy, Tamale, Ghana

References

[1] Abe S, Buri MM, Issaka RN, Kiepe P, Wakatsuki T. 2010: Soil Fertility Potential for Rice Production in West African Lowlands. JARQ. 44 (4), 343 – 355.

[2] Anzai T 1993: Effect of Succesive Applications of Organic Matter on the Cultivation of Paddy Rice in Gley Paddy Soil III. Influence of successive applications of rice straw ash on growth, yield and nutrient uptake of rice and on chemical properties of paddy soil. Bull. Chiba Agric. Exp. Stn. 34, 13-21.

[3] Austin LM, Duncker LC, Matsebe GN, Phasha MC, Cloete TE 2005: Ecological sanitation – Literature review. WRC Report No TT 246/05. WRC, Pretoria.

[4] Buri MM, Iassaka RN, Fujii H, Wakatsuki, T. 2010: Comparison of Soil Nutrient status of some Rice growing Environments in the major Agro-ecological zones of Ghana. International Journal of Food, Agriculture & Environment Vol. 8, No. 1, 384-388.

[5] Buri MM, Issaka RN, Wakatsuki T, Otoo E. 2004: Soil organic amendments and mineral fertilizers: Options for sustainable lowland rice production in the Forest agro-ecology of Ghana. Agriculture and Food Science Journal of Ghana, 3, 237-248.

[6] Cofie OO, Kranjac-Berisavljevic G Dreschel P 2005: The use of human waste for peri-urban agriculture in Northern Ghana. Renewable Agr. Food Sys, 20, 73-80.

[7] Food and Agriculture Organization 2004: Use of phosphate rocks for sustainable agriculture. FAO Fertilizer and Plant Nutrition Bulletin No.13, Food and Agriculture Organization of the United Nations, Rome.

[8] Ghana Seed Company (1988): Regional Branch Annual Report, Tamale.

[9] Issaka RN, Buri MM. Wakatsuki T. 2009: Effect of soil and water management practices on the growth and yield of rice in the forest agro-ecology of Ghana. J. Food Agric. Environ., 7 (1), 214-218.

14

Soil Fertility

[10] Issaka, R.N., Buri, M. M.. and Wakatsuki, T. 2009. Effect of soil and water management practices on the growth and yield of rice in the forest agro-ecology of Ghana. Journal of Food, Agriculture & Environment Vol.7 (1) : 214-218.

[11] JIRCAS 2010: The Study of Improvement of Soil Fertility with Use of Indigenous Resources in Rice Systems of sub-Sahara Africa. Business Report 2009.

[12] Lehmann J, Pereira da Silver Jr J, Steiner C, Nehls T, Zech W, Glaser, B 2003: Nutrient availability and leaching in an archaeological Anthrosols and a Ferralsol of the Central Amazon basin: fertilizer, manure and charcoal amendments. Plant Soil, 249, 343-357.

[13] Matsui S 1997: Nightsoil collection and treatment: Japanese practice and suggestions for sanitation of other areas in the globe. Sida Sanitation Workshop. Balingsholm, Sweden.

[14] Quarcoo AND 1996: A characterization and decomposition study of poultry manure. A BSc Thesis submitted to the University of Science & Technology, Kumasi Ghana.

[15] Senayah JK, Issaka RN, Dedzoe CD 2008: Characteristics of Major Lowland Rice-growing Soils in the Guinea Savanna Voltaian Basin of Ghana. Agriculture and Food Science Journal of Ghana, 6, 445-458.

[16] Yeboah E, Ofori P, Quansah G, Dugan E, Sohi S 2009: Improving soil productivity through biochar amendments to soils. African Journal of Environmental Science and Technology, 3.

Chapter 2

Soil Management for the Establishment of the Forage Legume Arachis pintoi as a Mean to Improve Soil Fertility of Native Pastures of Mexico

Braulio Valles-de la Mora,
Epigmenio Castillo-Gallegos,
Jesús Jarillo-Rodríguez and Eliazar Ocaña-Zavaleta

Additional information is available at the end of the chapter

1. Introduction

Pasture (rangelands) degradation in the humid tropics of Latin America is a fact that dates back several decades, and to date not only has not been resolved, but tends to worsen according to the unfavorable economic situation of livestock in the region [1].

In Mexico, according to a report [2], 75% of the degradation is caused by deforestation (25.8%), overgrazing (24.6%) and changing land use (agricultural and urban-industrial, 25.5%). The report adds that, in the north as well as in the southern of Mexico, livestock have overgrazed pastures and supports excesives stocking rates, causing a radical change in the floristic composition of rangelands and reduced permeability of the soil, increasing run-off and causes accelerated erosion thereof.

In this paper we addressed several land management practices for the establishment of the forage legume *Arachis pintoi* (CIAT accesions 17434, 18744 and 18748) as a means to improve soil fertility on native pastures of Mexico. *Arachis pintoi* was selected because it is a forage species that has enormous potential to improve the vegetation cover of the grazing areas in the Mexican tropics, and its contribution of nutrients to the soil, improving the fertility of this. All experiments were conducted in the northern of Veracruz state, Mexico, in a hot and humid climate, where soils are classified mainly as Ultisols or Oxisols. Some of the experiences were developed in native pastures and or in citrus plantations because this is a very important crop in this region.

In the most recent experience, three land preparation management experiments were conducted, in order to evaluate the establishment of *Arachis pintoi* CIAT 17434. The results offer a range of practices to cattle producers from which they could select the best practice according their specific conditions.

Previously, in 2006, two experiments were carried out in order to assess the establishment of *Arachis pintoi* as a cover crop in citrus plantations.

Also, were evaluated two treaments to establish *Arachis pintoi* and *Pueraria phaseoloides*. The two treatments consisted of (1) weeding by slashing (S) and application of herbicides (H), and (2) burning (+B) or not (-B), as main plots. Phosphorus (simple superphosphate) application (-P, +P) was included as subplots.

In other experiment, two methods of soil preparation were evaluated in a native pasture. The two methods, conventional tillage and minimum tillage were evaluated under the establishment of *A. pintoi* CIAT 17434 with fertilization (T1 and T2) and without fertilization (T3 and T4). In terms of number of *A. pintoi* plants established and soil cover, with complete soil preaparation, gave the best results. The legume did not respond to fertilization because of its slow initial growth.

2. Establishment of *Arachis pintoi* Krapov & W.C. Greg. as a cover crop in citrus plantations of Veracruz, México

The citrus crop in Mexico is one of the most important agricultural activities, both in area established as the value of marketing. At the end of 1999, 322.000 ha of orange (*Citrus sinensis* L.) and 32,000 ha of Persian lime (*Citrus latifolia* Tan), which depend altogether more than 15,000 families involved in the processes of production, harvesting, packing and marketing. In the case of the orange, the estimated production in 1997 was 3.9 million tonnes, while for the Persian lime was 244 thousand tons [3].

Arachis pintoi has shown a high potential as a cover crop in perennial crops such as citrus [4], peach [5], banana [6] and papaya [7]; thus, its incorporation in orange and lemon plantations in Mexico could be a viable alternative. Considering this, we evaluated the ecotypes *A. pintoi* CIAT 17434, 18744 and 18748 as options in citrus vegetation of Veracruz, Mexico.

2.1. Materials and methods

In April 1996 we established two experiments on commercial farms located in the municipalities of Martinez de la Torre and Misantla, Veracruz, Mexico (20 º 03 'north latitude and 97 º 03'longitud west), with hot and humid climate (24 º C average and 1980 mm annual rainfall), and no definite dry season, at 112-151 meters above sea level. Figure 1 presents the data of temperature and rainfall recorded during the course of the experiments, which is typical of the region, whereas data of 20 years, except for rain April, where the normal is half of that shown in graph mentioned.

2.2. First experiment

This was done in a lemon orchard Persian 3-year-old plantation trees with 7 x 7 m. We evaluated the establishment as cover crop of the ecotypes: CIAT 17434, 18744 and 18748. The field was prepared with cross harrowing, 20 cm deep. AP 17434 was used for vegetative material (stolons 20 cm in length) while the remaining were planted with seeds (two seeds per planting point). All ecotypes were inoculated with *Bradyrhizobium* strain CIAT 3101. Sowing was done in furrows separated by one meter and 50 cm between plants. For the availability of plant material or seed ecotypes had different number of test sites, with three repetitions of a square meter per site, so the 17434 had five sites, while the 18744, two, and the 18748, a site. The treatments were fertilized at planting date with superphosphate single at 50 kg/ha of P_2O_5, and 30, 90 and 180 days after planting date with KCl (83 kg/ha). Number of plants (plants/m2) and plant height (cm) was evaluated at 4, 8 and 12 weeks; and coverage (%) at 4, 8, 12, 16, 20 and 24 weeks. The data of number of plants were subjected to a logarithmic transformation to meet assumptions of analysis of variance. Analysis of variance were performed, and means were compared using the Tukey test. For plant height only averages were estimated. We used a completely randomized design, with *Ap* ecotypes as treatments, and a level of probability (P) for comparison of means of 0.05 was used.

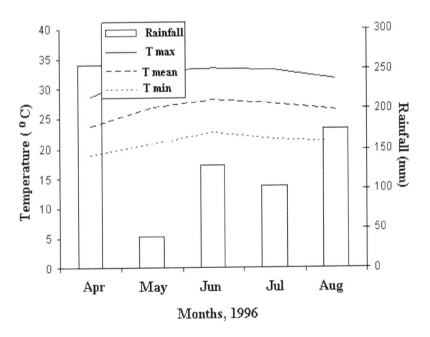

Figure 1. Climatic conditions during the establishment period.

In the case of coverage, the trend of the data indicated the existence of an asymptotic response, so exponential models were fitted to a maximum, logistic and sigmoid, using the routine "Regression Wizard" program SigmaPlot [8]. The model that final showed the best fit to the data coverage with rational values, was the three-parameter sigmoid, which is described below:

$$Y = a \bigg/ \left(1 + e^{-((X - X0)/b)} \right) \tag{1}$$

where "Y" is the coverage in percentage, at a "X" time, given in weeks; "a" is the maximum coverage value predicted by the model; "e" is the base of natural logarithms; "X0" is the time to "Y" reaches 50% of the value of "a"; and "b" is a constant of proportionality indicating the slope of the "S" on the right side (the higher the value, the greater slope), ie how fast it reaches the value of "a".

2.3. Second experiment

In this case, the orchard was located in the municipality of Misantla, Veracruz, and consisted of an orange plantation with coffee plants from 14 and 8 years old, respectively. The arrangement of citrus planting was 6 x 6 m, with four coffee plants around each orange tree. A week before the start of the experiment, the native vegetation was controlled with mechanical slashing and application of glyphosate (2 L/ha). The establishment of *Arachis pintoi* CIAT 17434 was evaluated for three methods of site preparation: disking, chiseling and hoeing, with two *A. pintoi* plant arrangements: plants at 35 and 50 cm within furrows separated by 75 cm; and two fertilizer levels: with and without P+K+Mg. P was used as triple superphosphate (50 kg/ha P_2O_5), for K, potassium chloride (50 kg/ha of K_2O), and Mg, magnesium sulfate (20 kg/ha) applied every 30 days post -planting. Plant material consisting of stolons of 20-25 cm was used, placing of 3-4 stolons per plant site.

A randomized blocks design was used, in an split-split plot arrangement, being fertilization treatment the main plot and subplot planting method, sub-divided into two planting densities. This resulted in 12 treatments with four replicates each. The total area was 2268 m^2, and the experimental unit was 144 m^2.

We measured the percentage of coverage, number of plants/m^2 and plant height (cm, five plants per replication), at 4, 8, 12, 16 and 20 weeks post planting. Coverage data, number of plants and plant height were subjected to analysis of variance, and means were compared using the Tukey test from the SAS statistical package [9]. The soil was analyzed at the beginning of the experiment and 16 months later to determine changes in organic matter, soil acidity, as well as levels of nitrogen, phosphorus and potassium. Economic estimates were made to determine costs of establishment, maintenance and return on investment, compared to traditional management of weed control in citrus plantations. Were considered: the cost of slashing of the land, legume plant material and its planting labor, fertilization (P, K, Mg), land preparation, with disking, hoeing; and herbicide application.

Soil Management for the Establishment of the Forage Legume Arachis pintoi as a Mean to
Improve Soil Fertility of Native Pastures of Mexico

19

2.4. Results

2.4.1. First experiment

*Number of plants.*Table 1 shows the average number of plants (and its standard error) for each accession. For the first and the second accesion an increase from week 4 to 12 was registered, while for the third accesion, the average remained constant during the period evaluated; achieving at 12 weeks an overall average of 5.1 plants/m^2. The analysis of variance did not detect any statistically significant difference within each accession, considering the weeks of sampling.

CIAT accesion	Weeks			
	4	8	12	P**
17434	2.5 ± 0.24*	3.1 ± 0.28	3.2 ± 0.33	0.0875
18744	3.8 ± 0.76	3.8 ± 0.51	4.2 ± 0.48	0.5833
18748	8.3 ± 0.33	7.7 ± 1.45	8 ± 1.0	0.8153

* Standard error of the mean.

** Probability level.

Table 1. Number of plants/m^2 from *Arachis pintoi* ecotypes established as cover crop in a citrus orchard soil of Veracruz, Mexico.

Plant height. Except for the evaluation at 4 weeks, the range was kept between 10 and 20 cm, and the latter value was more frequent in ecotypes 18744 and 18748.

Coverage. In the three ecotypes the model was highly significant ($P < 0.0001$), and in all cases the model parameters were different from zero at the same level of probability. R^2 values were greater than 0.8 (Table 2).

CIAT accesion	n	Model parameters ¥			
		a	X0	b	R^2
17434	90	95.0234	15.9488	3.9614	0.8332
18744	36	96.3046	12.3222	3.0102	0.9481
18748	18	94.1993	12.7762	3.4453	0.9105

¥ "Y" is the percentage of ground covered by the plant, "a" is the maximum coverage, "X0" is the time in weeks to reach half of "a", 'X' is the time in weeks, since planting date; and "b" is a constant of proportionality.

Table 2. Sigmoid model parameters: $Y = a/(1 + e^{-((X - X0)/b)})$, applied to the increase in coverage of three *Arachis pintoi* ecotypes, after planting.

In round numbers, weeks to reach 50% and 100% coverage were 16 and 32; 12 and 24; and 13 and 26, for ecotypes 17434, 18744 and 18748, respectively;

Showing the accesion 17434 the slowest establishment, considering that at 24 weeks, plants covered an average of 84%, compared to 18744 (94%) and 18748 (91%); these latter two, very close to the corresponding value of "a" (Table 2, Figure 2).

2.4.2. Second experiment

Number of plants and plant height. The information related to the number of plants and height of plants/m^2 are shown in Table 3. The disking treatment showed, on average, higher values for number of plants and plant height. This last parameter represented a range of 3.5 to 12.2 cm.

Coverage. Table 4 shows the percentages of coverage, achieved five months after establishment. In all treatments the highest values were achieved with the higher plant densities and fertilization treatment, except for planting treatments with hoeing. Treatments involving the disking had values far above the other ones, regardless of the plant density and/or fertilization applied. Analyses of variance performed within each site preparation, indicated statistically significant differences (P $\leq$ 0.05) considering the variables plant density and fe application or not of fertilizer.

Changes in the soil. In relation to changes in soil properties, increases were recorded for the content of nitrogen, phosphorus and potassium, although there was a decrease in organic matter content (Table 5).

Economic considerations. Economic estimates indicated that establishment costs per hectare (in U.S. dollars) for the year in which the experiment was performed, varied according to the evaluated treatments, being lower for those without fertilization (US $ 294, 410 and 396) in compared with those receiving fertilizer (US $ 356, 472 and 473) for treatments with disking, weeding and hoeing, respectively. Moreover, the expenses incurred to control weeds in one hectare included the purchase of a commercial herbicide (glyphosate), an adherent and implementation of both. It imported US $ 222.

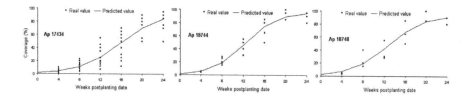

Figure 2. Dispersion data coverage and lines fixed to three-parameter sigmoid model, shown in Table 2.

2.5. Discussion

2.5.1. First experiment

The number of plants for the three ecotypes at 12 weeks, was on average lower compared with those found by [10] in one of three experiments with *Arachis pintoi* in native pastures of that region. He observed that Ap 17434 presented at that time more than 10 plants/m^2, using plant material grown in field where vegetation was controlled with a machete and herbicide, with or without burning the dead material and fertilized or not with P. The smaller number of plants found here could be due in part to the month after planting (May) was relatively dry (<50 mm), consequently affecting plant emergence. In the mentioned experiment [10] the growth period immediately to planting date had higher humidity (> 150 mm).

	Plants/m^2					Plant height (cm)				
Trat.	PD1+F	PD1-F	PD2+F	PD2-F	**P	PD1+F	PD1-F	PD2+F	PD2-F	P
D*	32.0	31.0	34.5	34.5	0.2641	11.6	8.2	12.0	8.0	0.5255
	(0.81)	(1.11)	(3.42)	(2.89)		(0.96)	(0.49)	(1.05)	(0.50)	
Ch	14.0	11.5	13.5	9.5	0.8672	4.6	5.0	5.6	3.5	0.0193
	(1.65)	(1.93)	(1.93)	(0.50)		(0.35)	(0.53)	(0.75)	(0.59)	
H	21.0	20.5	18.2	20.5	0.6755	9.5	10.2	8.3	10.9	0.2048
	(2.28)	(1.71)	(4.11)	(4.42)		(0.71)	(0.65)	(0.78)	(0.91)	

PD1 and PD2= Plants sown at 35 and 50 cm within the furrow respectively.

+F: With fertilization (kg/ha: 50 P_2O_5, 50 K_2O, 20 Mg_2SO_4); and, -F: without fertilization.

*D=Disking, Ch=Chiseling, and H=Hoeing.

**P: Probability level.

Table 3. Number of plants of *Arachis pintoi* and its height (Averages; standard error in parentheses), reached five months after planting date according to soil preparation, plant density (PD) and fertilizer application or not (F).

Moreover, it appears that the ecotypes evaluated here shown in the early stages of establishment a tendency of erect growth. It has been indicated [10] a range from 14.4 to 21.0 cm at 12 weeks post planting date, regardless of the treatments.

With respect to coverage, the R^2 values showed good predictive power for the environmental conditions during the study. No one model predicted a maximum coverage of 100%, because the measurement time was only 24 weeks.

In Colombia [4] evaluated in citrus plantations the same ecotypes, and found 8 months after that Ap 17434 was much lower coverage (32%) compared with the other ones (73% on average). On native pastures [10], found in another experiment with Ap 17434 that its establishment was even slower, since the accession planted with no-tillage or reduced tillage, with or without fertilization (P, K, Mg, Ca, Zn, Cu and B), needed 20 to 21 weeks to achieve 50% coverage.

The lower rate of coverage by the accession 17434 was also confirmed [11], on the experiment developed in this same region comparing four species of forage legumes (*Desmodium ovalifolium, Neonotonia wightii, Pueraria phaseoloides* and *Stizolobium deerigianum*) associated to a citrus plantation. This slowness in the establishment was also reported in Costa Rica [12] to associate in banana plantations.

Treatments	Weeks after planting					
Disking	4	8	12	16	20	P*
PD1+F	19.8±2.04	23.2±2.69	33.0±4.40	76.2±6.25	87.5±3.23	
PD1-F	10.5±1.04	17.7±3.17	23.7±10.5	52.5±10.5	70.0±7.90	
PD2+F	9.0±0.57	13.2±2.98	20.2±3.75	40.0±12.4	63.7±15.46	0.9641
PD2-F	8.0±0.71	10.5±1.26	10.5±3.23	26.2±3.14	52.5±9.11	
Chiseling						
PD1+F	6.0±0.91	4.2±1.31	7.0±1.29	10.2±1.11	11.7±1.08	
PD1-F	6.7±1.25	4.5±1.19	3.7±0.48	6.5±0.64	8.2±0.48	
PD2+F	6.2±0.47	6.5±0.29	5.7±1.31	7.2±1.43	9.0±1.78	0.0232
PD2-F	4.2±0.85	4.2±0.94	3.5±0.29	5.7±0.63	6.2±0.85	
Hoeing						
PD1+F	7.5±0.64	12.0±2.00	7.75±1.25	9.5±0.87	23.7±3.75	
PD1-F	8.7±1.10	8.7±1.89	8.7±0.47	20.0±4.56	33.7±5.54	
PD2+F	9.0±2.16	4.2±0.48	5.2±0.63	12.2±2.25	26.2±5.54	0.3202
PD2-F	8.5±0.50	5.0±0.71	10.0±2.38	14.0±3.81	27.5±7.22	

PD1 and PD2= Plants sown at 35 and 50 cm within the furrow respectively.

+F: With fertilization (kg/ha: 50 P_2O_5, 50 K_2O, 20 Mg_2SO_4); and, -F: without fertilization.

**P: Probability level.

Table 4. Coverage (%) of plants of *Arachis pintoi* (mean ± standard error) reached five months after planting date according to soil preparation, planting density (D) and the application or no fertilizer (F).

Soil factors	Start	End	Difference
Organic matter (%)	2.2	2.0	- 0.20
Nitrogen (Kg/ha)	8.9	28	+ 19.1
Phosphorus (Kg/ha)	6.0	40	+ 34
Potasium (Kg/ha)	86	301	+ 215
pH	5.0	5.8	+ 0.8

Table 5. Changes in the soil with the use of *Arachis pintoi* 16 months after planting.

Moreover, the ecotypes established by seed showed a higher rate of coverage, however, however, these differences in the velocity of establishing tend to disappear as time passes.

2.5.2. Second experiment

The coverage obtained with the disking treatment with plants every 35 cm along the furrow, and with or without fertilization are considered acceptable and are superior to those reported for *Ap* 18748 in coffee plantations of Nicaragua for high plant densities using vegetative material (strips of 3.3 m wide, with furrows 50 cm) and three weedings in the first 90 days [13]. This author reported that at 158 days post seeding, the legume exceeded 60% of ground cover. In Brazil [14], assessed *A. pintoi* at plant densities of 8 to 16 plants/linear m, reaching a 50% coverage to 84 and 68 days post seeding, respectively; whether the separation between furrows was 25 or 50 cm. The above percentages indicate superior performance under these conditions that found here, which is explained by the higher plant density used.

Respect to changes detected in the soil properties, the increase in the concentration of N could be attributed to a transfer to the soil of the element present in the leaves of *Arachis pintoi* by the effect of decomposition thereof. In this regard, [15] estimated litter decomposition of grasses and legumes, among whom was *A. pintoi*. They found that the decomposition of organic matter and nitrogen in leaves of this legume, along with that of *Stylosanthes capitata*, decomposed faster than the other species studied, although the amounts released of P, K, Ca and Mg were similar among grasses and legumes. Other researchers [13], working with *Ap* 18748 or *Desmodium ovalifolium* CIAT 350 associated with coffee plants, found no differences for any legume in N, P and K soil, three years after establishment; unlike [6], who in Australia, in banana plantation with or without *Arachis pintoi* after 5.5 years found significant increases in the association, in terms of organic matter (3.94 vs. 3.71%), N (0.42 vs. 0.39%); the K, Ca, Mg and Na increased to at 52, 26, 43 and 23%, respectively.

By comparing these costs with the traditional management of weed control in citrus orchards, we found that the costs for these plantations were around US $ 222 per year. Economic estimates in coffee plantations in Nicaragua [13], mentioned that the relative costs (%) in the establishment and maintenance of the associations were higher in the first two years, compared with the traditional control of weeds, but at that time the use of herbicides was lower between 30-50%. Establishment costs in the three experiments [see 10] fell in the range of US $ 282 to 623 (the exchange rate in 2001) in terms of inputs applied. Although costs for the establishment of *Arachis pintoi* is higher, this is recovered in about a year and a half or two, with the advantage of having a highly competitive species for weed control and its long persistence in the land, plus inputs of nutrients to the soil as an additional benefit.

2.6. Conclusions

Arachis pintoi is a promising legume to associate as a cover crop with citrus plantations and other crops of high commercial value, such as bananas, pineapple, coffee and papaya. In the case of the first experiment, *Ap* ecotypes CIAT 18744 and 18748 represent for citrus plantations area of Veracruz, a better option compared to *Ap* CIAT 17434, due to the slowness of this accesion to cover the ground. Regarding the second experiment, the disking treatment, proved to be the best treatment for the establishment of the legume, but the costs of establishment will vary depending on the inputs applied, but a long-term coverage will absorb these costs converting this costs in an effective alternative.

3. Establishment of *Arachis pintoi* in native pastures of Mexico

Research results from the hot humid areas of México and from other parts of Latin America showed that the forage legume *Arachis pintoi* CIAT 17434 has the ability to be associated with grasses, because it has shows better persistence than other legumes and also has high nutritive value and palatability [16, 17-20]. *A. pintoi* establishment techniques range from a complete soil tillage and planting with seed to zero tillage and planting with vegetative material (stolons) into an existing pasture [17]. The objective of this study was to evaluate the agronomic performance of different techniques of establishing *A. pintoi* CIAT 17434, as well as the accessions CIAT 18744 and 18748, into existing native pastures in the humid tropics of the coastal plains of the Gulf of México.

3.1. Materials and methods

3.1.1. Site characteristics

Three experiments were conducted during 1991 and 1996 at the Centre for Teaching, Research and Extension in Tropical Animal Husbandry (CEIEGT, its acronym in Spanish) of the Faculty of Veterinary Medicine, of the National University of Mexico (UNAM). The Centre (CEIEGT) is located in the eastern coastal plain of México about 40 km West of the Gulf of México coast line at 20° 02' N and 97° 06' W, at 112 m a. s. l.

The climate is hot and humid, with rains all year round. Mean yearly rainfall was 1,917±356 mm from 1980 to 1997. Monthly rainfall is highly variable being September (322 mm) and October (248 mm) the rainiest months while March (85 mm) is the driest. The coldest and hottest months are January (18.9 °C) and June (27.8 °C). Minimum daily temperatures from November to February (winter) are around the critical range of 8-10 °C, below which the growth of C_4 tropical grasses is severely reduced [21-23]. These combinations of rainfall and temperature lead to a seasonal DM production pattern, a common situation in the tropics of Latin America: A high growth rate on the rainy season followed by poor growth during the winter and dry seasons.

The experiments were conducted in different years. Temperatures were typical of each season, but the current maxima were below, and the current minima above the long term (1980-1997) mean (Figure 3a). Total rainfall during experiment 1, December 1991 to September 1992, was 39% above average (Figure 3b). Rainfall in the experimental planting seasons was 339 mm in winter (November 29, 1991 to February 14, 1992), 637 mm in the dry season (March 2 to May 18 of 1992) and 1,352 mm in the rainy season (July 2 to September 17 of

1992). Rainfall was 19% above average during experiment 2 in 1993, but rains in 1996 were 43% below average for experiment 3 (Figure 3b).

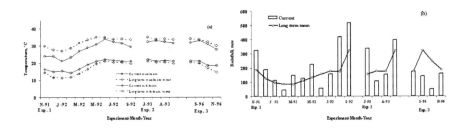

Figure 3. Current and long term monthly temperatures (a) and rainfall (b) for the 3 experiments.

The soils are acid Ultisols (Durustults), with a range in pH from 4.1 to 5.2, and an impermeable hardpan between 0 and 25 cm in depth, that result in a inadequate drainage during the rainy and winter seasons. The soil texture is clay-loam with low levels of P (< 3 ppm), S (< 30 ppm), Ca (< 3 meq/100 g) y K (< 0.2 meq/100 g). Both cation exchange capacity and aluminum saturation increase with depth, but the latter do not reach toxic levels for pasture plants [24].

3.1.2. Experiment 1. Reduced and zero tillage, with or without fertilisation

The study was conducted to test the combined effects of tillage type: reduced and zero, and fertilisation with (kg/ha): P 22; S 25; K 18, Mg 20; Ca 100; Zn 3; Cu 2 and B 1, or no fertilisation, in a four treatment combination: T1, reduced tillage and fertilisation; T2, reduced tillage without fertilisation; T3, zero tillage and fertilisation; and T4, zero tillage without fertilisation. Reduced tillage consisted of four passes of a disk harrow, while zero tillage only required the elimination of pasture vegetation by machete to ground level.

The experimental area was 2,000 m² (50 m x 40 m split in two plots of 1,000 m² - 25 m x 40 m). These plots were divided in two sub plots of 500 m² (25 m x 20 m), of which one sub plot was fertilised. Three 2,000 m²-experimental areas were used: one per each climatic season (winter, dry and rainy season).

Arachis pintoi was planted on sub-plots of 500 m² on 29 November, 1991 (winter season), 2 March, 1992 (dry season) and 2 July, 1992 (rainy season). Three to four stolons, approximately 15 cm in length and with five nodes per stolon, were planted per planting position. On the reduced tillage treatments the distance between rows and planting positions were 1.0 m and 0.5 m, respectively. Planting was done on 3 m wide strips, which alternated with 3 m intact native pasture strips. Three rows of the legume were planted per strip and 3 strips were contained in a subplot, being the sampling quadrat size 3.0 m x 1.5 m. On the zero tillage treatment, distance between rows and positions was 2 m and 0.5 m, respectively, with the subplot containing nine sampling rows also and a sampling quadrat dimensions of 6 m x 3 m. Even though this planting arrangement was confounded with tillage treatments, it gave a similar number of planting positions per sub-plot and two sampling hills/m^2 in each sampling quadrat, regardless of type of tillage. Fertiliser was broadcast 30 days after planting.

3.1.3. Experiment 2. Type of control of native pasture growth, with or without P fertiliser

This experiment tested the combined effect of the type of pasture vegetation control: herbicide (glyphosate) or slashing (by machete) with or without burning of dead vegetation, and with or without localised P-fertilisation which resulted in eight treatment combinations. The choice of treatments attempted to reduce competition to *A. pintoi* from existing native pasture vegetation, enhance legume establishment and early growth, following the approach described by [25] for the establishment of legumes into existing Speargrass (*Heteropogon contortus*) native pastures, in Australia.

Slashing was done by machete and burning was carried out between 1-5 days after slashing. A 2% aqueous solution of glyphosate (480 g of isopropyl amine salt of glyphosate/l) was applied on a 0.25 m wide strip 15 days before planting; burning was done 15 days after herbicide application.

The planting legume was done between June 28 and July 3. Application of herbicide and herbicide plus burning, and slashing or slashing plus burning, were applied 15-16 days and 3-5 days earlier, respectively. Vegetative material, 0.25 m length stolons with eight nodes, was used for planting. This material was inoculated just prior to planting with a specific *Bradyrhizobium* culture obtained by suspension of 1 kg of profusely nodulated *A. pintoi* ground roots in a solution 7.5 litres of water and 1.5 litres of sugarcane molasses. Three stolons per planting position were put in a hole and covered with soil, allowing about 1/3 of the stolon to remain above ground. Distances among rows and planting position were 1.0 m and 0.5 m, respectively. The sub plot (10.0 m x 6.5 m) had 10 rows with 14 planting positions/row. Two sampling quadrats (2 m x 1 m) each with 4 planting positions were randomly allocated per sub plot. Single super phosphate (30 kg of P/ha) was applied at planting in a 0.07 m depth hole adjacent to the planting position.

3.1.4. Experiment 3. Establisment of Arachis pintoi accessions using seed pods

This experiment compared the establishment of three *A. pintoi* accessions using seed pods: CIAT 17434 (cv. Pinto peanut or Amarillo), 18744 and 18748. Seed germination was assessed in the laboratory at room temperature; using 125 seeds per accession. Petri dishes, bottom-lined with filter paper, were used and were watered twice daily. The seeding rate was equivalent to 10 kg of germinable seed pods per hectare, based on quadruplicate germination tests. The experimental plots (10 m x 5 m; ten 5 m length rows/plot) were established within a grazing experiment where milk production from native pastures and native pastures associated with *A. pintoi* was to be compared. Three replicates were established in one paddock and three in another. Each replicate had three plots, with an accession each. Plots were excluded from grazing for the 12 weeks of the establishment period. A 2% aqueous solution of glyphosate was applied on a 0.30 m wide strip 15 days before planting to eliminate competition from existing vegetation. Distance between rows and planting positions was 1.0

m and 0.5 m, respectively. Seed pods were placed in a 5 cm deep hole made with pointed wooden stick, and lightly covered with soil by the planter's foot. Three replicates were planted on August 2 and three on September 3, 1996. Fertiliser was not applied.

3.2. Measurements and statistical analyses

The response variables were: 1) plant number (PN, plants/m^2) by counting; 2) plant height (PH, cm), on each plant within the sampling quadrat, measured with a ruler from the soil surface to the uppermost part of the plant; and 3) soil covered by the legume or cover (COV, % of quadrat area covered by the legume) measured with the aid of a 1 m^2 quadrat, divided into 25 squares, which was placed over the row. These measurements were done on weeks 4, 8 and 12 after planting [26]. In experiment 1, PH was not measured, but COV was measured again at 24 weeks after planting.

In experiment 1, there were no field replications, since it was perceived that treatments applied in larger areas would have a closer resemblance to that of farmers' fields. Also, if several sampling quadrats were used within each treatment plot, this would yield information as useful as that obtained from randomised complete block designs. In experiments 2 and 3, the design was a randomised complete block design with 3 blocks as replicates. The treatment arrangement was a split-plot in experiment 2, where the main plot was the combination of type of pasture vegetation control (slashing and herbicide), while the combinations of burning (with and without) and P application (with and without) were the sub-plots; additionally the effect of time after planting was considered a sub-sub-plot. The treatment arrangement of the third experiment was a split plot, in which the main factor was the combination of month of planting by accession and time after planting the sub-plot. Here, number of plants was expressed as "plants/50 m^2", in order to be clearer and avoid fractions of plant/m^2. Analyses of variance were done with linear additive models in accordance to the experimental design [27]. The natural log transformation of the response variable was used if its response to time was exponential. If necessary, linear or exponential relationships provided rates of increase with time in the measured variables. Also, means comparisons using Tukey's test were done when was necessary.

3.3. Results

3.3.1. Experiment 1. Reduced and zero tillage, with or without fertilisation

The main effect of treatment on plant number (PN) was highly significant (P<0.01) in all seasons. The linear effect of week after planting was highly significant (P<0.01) on PN in the winter season of 1991-92 and the rainy season of 1992, but it was not significant (P>0.05) in the dry season of 1992 (Table 6). There was no significant treatment x week interaction on PN in any season. The main effects of treatment and week after planting and its interaction were highly significant (P<0.01) on COV, except for the interaction in the rainy season. Weeks to reach 50% cover were 21, for T2 (winter season) and T4 (dry season); and 20, for T1 and T4 in the rainy season (Table 7).

	Treatments		Season		
	Tillage	Fertilisation	Winter	Dry	Rainy
T1:	Reduced	With	$1.36^b \pm 0.06$	$0.78^a \pm 0.06$	$2.56^a \pm 0.17$
T2	Reduced	Without	$1.70^a \pm 0.09$	$0.73^{ab} \pm 0.07$	$2.56^a \pm 0.22$
T3	Zero	With	$0.81^c \pm 0.03$	$0.60^b \pm 0.02$	$0.96^b \pm 0.09$
T4	Zero	Without	$0.82^c \pm 0.04$	$0.59^b \pm 0.03$	$0.80^b \pm 0.07$
	Effect of week after planting:		$1.16^{**} \pm 0.04$	$0.68^{NS} \pm 0.02$	$1.72^{**} \pm 0.10$

$P \leq 0.0001$.

Table 6. Effect of treatments on the number of plants of *Arachis pintoi* CIAT 17434 (pl/m^2) by season (Mean ± standard error), according to the tillage by fertilisation combination in experiment 1.

	Treatments		Season		
	Tillage	Fertilisation	Winter	Dry	Rainy
T1	Reduced	With	22 ± 0.4	25 ± 0.8	20 ± 0.3
T2	Reduced	Without	21 ± 0.3	24 ± 0.7	21 ± 0.5
T3	Zero	With	23 ± 0.4	23 ± 0.5	21 ± 0.4
T4	Zero	Without	24 ± 0.3	21 ± 0.6	20 ± 0.2

$P \leq 0.01$.

Table 7. Mean ± standard error for weeks to reach 50% cover by *A. pintoi* CIAT 17434 according to the tillage by fertilisation combination in experiment 1.

3.3.2. Experiment 2. Control of native pasture growth, with or without P fertiliser

The effect of time after planting was highly significant ($P<0.01$) upon all response variables. Height values increased with time, but to a different degree on each main plot combination. The increase in plant height (PH) with time was much larger than the increases with time shown by the other two response variables. The standard deviations were high in all cases and increased with time also (Figure 4). The coefficients of variation remained relatively uniform through time: 28% to 31% for plant number (PN), 29% to 35% for plant height, and 75% to 83% for cover (COV).

When herbicide was applied, the burned plots produced taller plants than the non-burned ones ($P= 0.01$), but the contrary happened on slashed plots ($P<0.05$) (Table 8).

P fertilisation did not increase ($P>0.05$) legume cover in any vegetation control by burning combination. Slashing without burning and without fertiliser, the treatment with the least external inputs, had significantly ($P<0.05$) less legume cover than the herbicide plus burning plus fertilisation treatment, the treatment requiring the most external inputs (Table 9).

Soil Management for the Establishment of the Forage Legume Arachis pintoi as a Mean to
Improve Soil Fertility of Native Pastures of Mexico

29

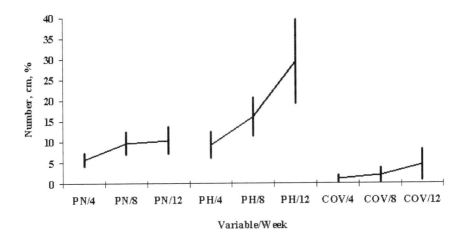

Figure 4. Effect of time after planting (4, 8 and 12 weeks) on *A. pintoi* CIAT 17434 plant number (PN, number/m²),
plant height (PH, cm) and legume cover (COV, %). The vertical lines are the standard deviations.

Treatments		Plant height (cm)	Statistical significance of the non-burning vs burning comparison within vegetation control
Vegetation control	Burning		
Herbicide	Without	14.54 ± 1.14	0.01
Herbicide	With	21.01 ± 1.57	
Slashing	Without	20.89 ± 1.23	0.05
Slashing	With	17.09 ± 1.25	

Table 8. Combined effect of vegetation control x burning treatments upon *A. pintoi* CIAT 17434 mean plant height
(PHT, cm).

3.3.3. Experiment 3. Establishment of three A. pintoi accessions using seed pods

The averages of percentage of seed germination at 7 days on the laboratory were of
44.8±4.08, 44.8±4.45 and 32.8±1.50, for CIAT 17434, CIAT 18744 and CIAT 18748, respectively; and values (percentages) of emergence at 7 days after planting were 91.3±1.5, 82.0±2.4
and 73.8±1.4, respectively, which were statistically different among them (P<0.05). The main
effects of month of planting and accession were significant (P<0.05) on COV. Legume cover
increased linearly with time (4, 8 and 12 weeks), but without differences in slope among accessions (Figure 5). Using the regression equations of cover *vs*. time, it was calculated that

for the August planting it took 45, 46 and 56 days for accessions CIAT 17434, CIAT 18744 and CIAT 18748, to cover 5% of the soil, respectively. Values for September were 55, 50 and 55 days. Plant height was affected by month of planting (P<0.01), the plants being taller in August. The interaction month x accession was significant (P<0.05), but the accession CIAT 17434 was about 2 cm shorter than the others in both planting months (Table 10). Maximum height at the end of the establishment period was greater for August (27.4 cm) than for September (18.2 cm).

Treatment combination			Cover, %
Vegetation control	Burning	P-fertilisation	
Herbicide	Without	Without	2.39 ± 0.45
		With	2.18 ± 0.54
	With	Without	2.48 ± 0.42
		With	4.21 ± 0.91
Slashing	Without	Without	1.74 ± 0.31
		With	2.59 ± 0.84
	With	Without	3.17 ± 0.69
		With	2.01 ± 0.52

P≤0.01.

Table 9. Combined effect of vegetation control x burning x P-fertilization treatments upon *A. pintoi* CIAT 17434 mean cover (COV, %, ± standard error).

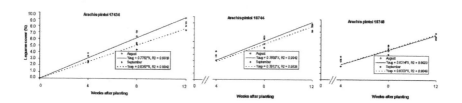

Figure 5. Effect of time after planting (4, 8 and 12 weeks) upon soil covering (%) of three *A. pintoi* accessions in experiment 3.

3.4. Discussion

In experiment 1, reduced tillage gave better results than zero tillage during the winter season, but the opposite occurred in the dry season. As soil moisture and temperature conditions increased in the rainy season, the difference between reduced and zero tillage not disappeared and was significant. Other trials conducted in the same region have indicated the advantage of reduced tillage over zero tillage to establish vegetatively *A. pintoi* [28]. The literature shows a general agreement among researchers in that some sort of soil disturbance is necessary to assure establishment [29, 30].

Month	CIAT accession	Cover, %	Plant Number	Plant height, cm
	17434	6.4 ± 0.8 [a*]	109 ± 5 [a]	11.9 ± 1.4 [b]
August	18744	6.2 ± 0.8 [a]	106 ± 6 [a]	13.8 ± 1.6 [ab]
	18748	5.1 ± 0.7 [b]	97 ± 6 [b]	14.3 ± 1.6 [a]
		5.9±0.42	103.8±3.29	13.3±0.87
	17434	5.1 ± 0.7 [NS]	124 ± 2 [a]	9.6 ± 0.9 [b]
September	18744	5.8 ± 0.7 [NS]	113 ± 3 [ab]	12.1 ± 1.0 [a]
	18748	5.2 ± 0.6 [NS]	99 ± 4 [b]	10.0 ± 0.9 [b]
		5.4±0.38	112.0±2.61	10.6±0.55

Means followed by the same letter are not statistically different at P≤0.01.

NS= Non-significant.

Table 10. Mean ± standard error of cover (COV, %), plant number (PN, plants/50 m^2) and plant height (PH, cm) per month of planting by accession combination in experiment 3.

It has been suggested [25] that seedlings facing more root competition from existing vegetation responded to fertilisation, whereas those without competition had a lesser or nil response.

In the winter season planting of experiment 1, fertilisation failed to stimulate COV of slashed plots, those supposedly with a larger competition from existing pasture. In the dry season planting, fertilisation was detrimental to COV in the slashed plots, in contrast to [25]; finally, in the rainy season the effect of fertilisation was negligible. The second experiment showed a positive effect of fertilisation on COV only when herbicide was applied and the dried vegetation was burned. When plots were slashed, but not burned, the effect of fertilisation on COV was positive. Nevertheless, when the slashed plots were burned, the fertilisation effect on COV was negative.

Fertilisation with 23 kg P/ha, 25 kg K/ha, 20 kg S/ha and 20 kg Mg/ha had a positive effect on COV (83.4% *vs.* 61.3%) and PH (12.0 cm vs. 8.6 cm) when the soil was prepared with 4 passes of disc harrow, but with zero tillage, fertilisation reduced both COV (25.0% vs. 30.6%) and PH (8.4 cm vs. 10.1 cm) [28].

As suggested by the inconsistent results of our trials and those of the literature, fertilisation appears not to be of great importance for the establishment of *A. pintoi*, when vegetative material is used. The lack of P response on *Arachis* species has been reported by other researchers. In experiment 2, single superphosphate was used, and perhaps the use of this source could explain, partially, the lack of response. Also, the very low P levels on soils at CEIEGT

(0.6 to 1.2 µg g^{-1} soil on 0-30 cm depth), could limit N mineralization [31], resulting in a poor legume performance.

In experiment 2, burning was directed to reduce competition from existing grasses, since the way *A. pintoi* vegetative material was planted assured a close contact with the soil. However, burning, as well as fertilisation, did not show a clear positive trend either on COV or on PHT.

When only herbicide was applied in bands in experiment 2, pasture canopy height was not reduced, leading to reduced PH of *A. pintoi*. On the other hand, when the herbicide treated vegetation was burned, PH of *A. pintoi* was not impeded. Non-burned plots gave slightly taller *A. pintoi* plants than those burned. *A. pintoi* CIAT 18744 flowers less and produces a denser stolon mat than the other two accessions and it also has a vigorous initial growth, covering the soil more rapidly than the CIAT 17434 accession [32-33]. For this reason, a better behaviour during establishment, particularly with respect to COV and PN was expected from this cultivar. Nevertheless, in experiment 3, COV performance at the end of establishment was similar to that of CIAT 17434 (8.5% *vs.* 8.7%) and only slightly better than CIAT 18748 (7.5%). Then, the 3 accessions behaved similarly during establishment. Rates of plant emergence are considered to be good, as *A. pintoi* is a legume that can have a strong dormancy [34]. However, emergence (from 125 seeds originally planted/plot) of new branched plants/plot was not so bad, considering that these values ranged from 70% to 90% for three accessions. Therefore, there was low coverage but high number of new branched plants. This situation is common for *A. pintoi*, which is characterized by its slow establishment, as has been reported [6, 35-36]. Zero tillage failed to stimulate a rapid establishment of *A. pintoi* in these trials, the reproductive mechanisms of this species ensure that eventually it will establish and encroach within the pasture. Our experience with this legume is that eventually it ends up to be the dominant species when associated with native pasture, Stargrass, or to both. A good strategy would be to establish *A. pintoi* in strips with reduced tillage at high density. This will result in a rapid establishment of a mixed sward in a minimum of time.

3.5. Conclusions

Neither fertilisation nor burning were successful in enhancing *A. pintoi* establishment; slashing did not improve establishment either. On the contrary, herbicides were effective and improved establishment over slashing. The best alternative to introduce *A. pintoi* into a native pasture is by reduced soil tillage in strips using, within the strips, 8 kg of pure live seed pods/ha; or 0.70 m between rows and 0.35 m between planting positions for vegetative material.

4. Establishment of *Arachis pintoi* CIAT 17434 and *Pueraria phaseoloides* CIAT 9900 using minimum tillage in Veracruz, Mexico

In the watershed Gulf of Mexico region, there is a highly seasonal pasture production due to climate variability. The main genera are components of *Paspalum*, *Panicum* and *Cynodon*

(Gramineae), and in smaller proportions *Centrosema* and *Desmodium* [37]. Among the legumes evaluated in that area, *A. pintoi* CIAT 1434 and *Pueraria phaseoloides* CIAT 9900 outstanding for their performance and good adaptation [38].

The cost of establishing pastures in native savanna vegetation is high when following traditional methods. Given this, it is justified to evaluate planting systems cheaper, to promote the adoption of new forages and their use to recover degraded pasture [39]. Therefore, this trial is performed to supporting evidence to assess the effect of various types of tillage and application of phosphorus on the establishment of *A. pintoi* CIAT 17434 and *Pueraria phaseoloides* CIAT 9900 in native pastures.

4.1. Materials and methods

4.1.1. Characteristics of the experimental site

The research was conducted at the Centre for Teaching, Research and Extension in Tropical Animal Husbandry of the Faculty of Veterinary Medicine, of the National University of Mexico (UNAM), located in north-central region State of Veracruz, 20 ° 4 'north longitude 97 ° 3' W and a height of 105 meters above sea level. The climate is hot and humid with rain all year, type Af (m) with average daily temperature of 23.4 ° C and average annual total precipitation of 1840 mm (1980-1989). The soil texture ranges from sandy loam to sandy clay. The area has a hard horizon with low permeability that occurs between 5 and 25 cm deep. The soils are acidic (pH 4.1 to 5.2), and are classified as Ultisols.

We used an area of 6.000 m^2 of degraded native pasture grazed by cattle. The treatments were the type of weeding (slashing, S; and herbicide, H) and the burning (B) application or not (with + B and without -B), to temporarily control the growth of existing vegetation (larger plots), and thus prove its effectiveness to allow the establishment of the legumes *Arachis pintoi* CIA 17434 (Ap) and *Pueraria phaseoloides* CIAT 9900 (Pp). Additionally we evaluated the application of phosphate fertilizer or not (+P addition; no-P as single superphosphate). The factorial combination between legumes and fertilizer was the subplot.

Treatments were applied between 28 May and 3 June 1993. The slashing (S) was a machete to the whole plot. In S + B, the burning was applied between one and five days after slashing. The application of herbicide (H) was done using a backpack sprayer, applied in bands 50 cm wide, spaced 1 m apart from the center of each. The dose was 0.96 kg (2 l) of a nonselective systemic herbicide (Glyphosate). The product was dissolved in 200 l of water and applied 15 days before seeding. In H + B, herbicide application was the same way as above, burning 15 days after application, only the bands where the herbicide is applied.

Ap vegetative material, was inoculated with the specific *Rhizobium*, by means of a suspension prepared with nodulated roots, washed and crushed to release *Rhizobium* bacteria, then adding cold water and molasses, placing the suspension in a refrigerator, performing all procedure in the shade. Each kg of root was added 1.5 kg of molasses (as adherent) and 7.5 l of cold water.

Legumes were planted between 3 and 5 days after applying treatments S or S + B, and between 15 and 16 days after applying treatments H or H + B. *Ap* was planted with stolons of 20 cm long, inoculated with the suspension of *Rhizobium* already described. By planting, we used a metal digging stick, to make a biased hole of 15 cm lenght and 5 cm depth. Three stolons were placed by hole and soil was compacted with foot to ensure contact with the ground. The distance between plants and rows was 0.5 and 1 m, respectively.

For planting of kudzu (*Pp*) botanical seed was used, previously scarified with sulfuric acid to 98% for 10 minutes. This ensured the seed germination in three days post seeding. The effectiveness of this treatment has been verified by other researchers [40].

Planting density was 2 kg/ha of pure and viable seed. After scarified, the seed was inoculated and seeded similarly as *Ap* placing about 8 seeds per site, but was not covered preventing soil compaction. The distance between plants and rows was 0.5 and 1 m, respectively. Single superphosphate (333 kg/ha = 30 kg P/ha) was applied at planting time in 5 cm band from the seed or plant material.

We used two sampling sites per plot at random. Firstly, two rows of each plot were chosen, and then the sampling site within each row. The recommended [26] variables were, number of plants, plant height (cm) and coverage (%) at 4, 8 and 12 weeks post seeding. A randomized complete block design was used, with three replications and a split-plot arrangement, with the factorial combination between weeding and burning as main plots, and the combination of the two legumes with or without fertilization as subplots. We considered the costs for materials and labor costs per treatment.

4.2. Results and discussion

Climate. – The climatic parameters of precipitation and temperature were recorded from May to September 1993. The monthly average temperature was similar for the periods, ranging from 25.5 to 27.0 °C. The lowest rainfall occurred in July and highest in September with 109 and 360 mm respectively. Rainfall totaled 1257 mm. This caused flooding which affected the establishment of each legume.

Number of plants. - Analysis of variance indicated that there was a highly significant effect ($P < 0.01$) of the species, with 1.77 and 0.55 for *Ap* plants/m² plants / m² for *Pp*. The number of plants for *Ap* can be considered acceptable, even expected 2 plants/m². Flooding caused by high rainfall brought rot of stolons.

The small number of *Pp* plants is also attributed to the seed rot because of soil waterlogging. It has been mentioned [41] that heavy rainfall limits the development of Kudzu (*Pp*). Also, surprisingly, the number of plants decreased as time passes ($P < 0.01$): There were 1.27, 1.18 and 1.0 plants/m² for first, second and third samples. Effects such as slashing, burning, and their interaction were not significant ($P > 0.05$), which coincides with other experiment [42]. These authors, who established three species of legumes (*Centrosema pubescens, Macroptilium atropurpureum* and *clitoria ternatea*) using total soil preparation, harrowing, plowing and burning, with no significant difference found ($P > 0.05$) among the different methods,and concluded that burning favored the establishment of legumes.

Plant height. - Analysis of variance showed highly significant differences (P <0.01) between the species: *Ap* with 18.3 cm and 9.5 cm with *Pp* (Figure 6). This difference is attributed to *Ap* was seeded with plant material starting its growth as seedling, which gave to Ap an advantage over *Pp* that was sown with seed.

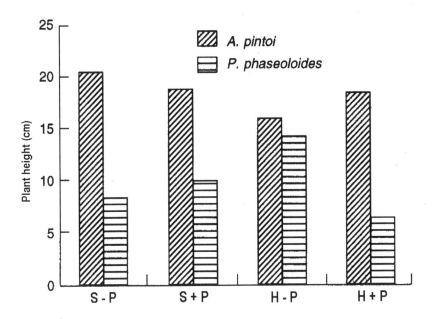

Figure 6. Effect of type of control weeds (herbicide –H- and slashing –S-), with and without P fertilizer over plant height of *Arachis pintoi* and *Pueraria phaseoloides.*

The interaction slashing X burning was highly significant (P <0.01). The application of H+B resulted in greater plant height with 16.0 cm, followed by S-B with 15.6 cm, being H-B method the lowest height with 11.1 cm. In the case of H+B, the plant height was attributed to no competition between the legume and native grasses; also, burning causes release of soil nutrients that legumes can absorb quickly, making their establishment more effectively.

Burning, releases mineral nutrients immobilized in plant tissues, and others are transformed into simple soluble salts, readily available to the plant [43]. In the treatment of S-B, the largest plant height was mainly due to competition for sunlight by the grass. Competition for sunlight between *Aeschynomene*, seeded with the grass *Hemarthria altissi-*

ma resulted in greater height during legume establishment [44]. Here, the combination of H-B produced lesser height.

Sampling at 4, 8 and 12 weeks showed highly significant differences (P <0.01) with 7.10, 12.23 and 22.39 cm, respectively (Figure 7). This increase in plant height in time was an expected effect.

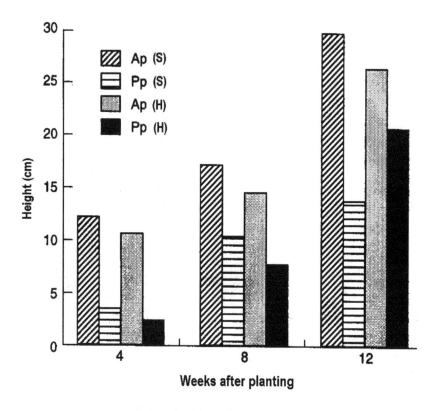

Figure 7. Effect of type of control weeds (herbicide –H- and slashing –S-), durin the three climatic evaluation seasons over the plant height of *Arachis pintoi* (Ap) and *Pueraria phaseoloides* (Pp).

The interactions species x weeding x fertilization, and weeding x species x sampling were significant (P <0.05), while weeding x fertilization x species x sampling were highly significant (P <0.01). Most probably is that the latter would have been highly significant because it contained the first two.

These results coincide with those of an experiment in Cuba [45], who evaluated different methods of control vegetation during the establishment of *Leucana leucocephala*, and reported that the best method was the application of systemic herbicide, achieving a plant height of 162.5 m and plant coverage of 96% to 5 months post seeding, concluding that the best promoter of the successful establishment of this legume was the control of vegetation.

Coverage. - The analysis of variance showed highly significant differences (P <0.01) between species: *Ap* showed a coverage of 2.6% and 0.5% Kudzu. We also found highly significant difference (P <0.01) between samples, with 0.7, 1.2 and 2.8% at 4, 8 and 12 weeks post seeding. The species x sampling interaction was highly significant (P <0.01). *Ap* was the best species, averaging 1.2, 2.0 and 4.6% while *Pp* averaged 0.2, 0.5 and 0.9% for the same samples (Figure 8).

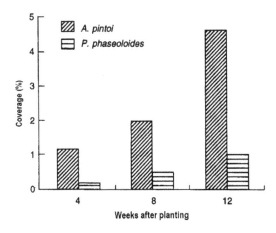

Figure 8. Soil coverage (%) according to the total experimental area, by *Arachis pintoi* (Ap) and Pueraria phaseoloides (Pp) at 4, 8 and 12 weeks afetr planting date.

The interaction slashing x fertilization x burning was also significant (P <0.05), resulting in the best combination of the H+B+P with 2.5% coverage, followed by S-B-P with 1.9%. Burning + fertilization promoted a good establishment of legumes. The combinations in which was planted after herbicide application, showed no significant differences for the variable coverage.

The burning x slashing x sampling interaction was also significant (P 0 <0.05). In the third sampling, treatment H+B+P was the best combination of coverage averaging 4.2%, followed by H-B-P with 2.4%. The other combinations were not significantly different from each other. The combination S+B+P coverage reached 3.7%, which is the highest value, which shows that the burning had positive influence on legume development, although interacted differently with the type of weeding and fertilizing.

The elimination of competition below and above ground, by applying H+B+P promotes the successful establishment of legumes. The lack of competition, plus the application of P, allowed to establish successfully the legume Siratro (*Macroptilium atropurpureum*) on the native grass [25].

The higher cost of treatment to establish *Ap*, was the S+B+P, or S+B-P (USD $ 195.00/ha), whereas the application of H-B-P was less expensive to establish *Pp* (USD $ 86.00/ha). Herbicide application was more economical compared to the slashing treatment. (Table 11).

4.3. Conclusions

The banded herbicide application without application of fertilizer is the best method for introducing vegetatively Ap in native grass pastures in north-central region of Veracruz State, Mexico.

Treatments				Cost (USD/ha)
Slashing	-B	- P	A. pintoi	173.43
			P. phaseoloides	117.16
		+ P	A. pintoi	193.43
			P. phaseoloides	137.15
	+ B	- P	A. pintoi	175.85
			P. phaseoloides	118.12
		+ P	A. pintoi	194.80
			P. phaseoloides	138.12
Herbicide	- B	- P	A. pintoi	142.15
			P. phaseoloides	85.57
		+ P	A. pintoi	162.15
			P. phaseoloides	105.57
	+ B	- P	A. pintoi	143.21
			P. phaseoloides	86.53
		+ P	A. pintoi	106.53

Table 11. Economic costs of treatments on the establishment of *Arachis pintoi* and Pueraria phaseoloides. Mexico, August 1996.

5. Establishment of *Arachis pintoi* CIAT 17434 by two tillage methods in a native pasture of Veracruz, Mexico

In the humid tropics of Mexico, native pastures are affected by climatic variations from one season to another that make it difficult, to obtain stable yields of forage during the year.

Also, financial constraints of most producers in the tropics must be considered when trying to introduce forage species [40]; so it is justified, evaluate and implement systems-on planting native vegetation, different from the traditional, in order to encourage the adoption of new and improved grass species, the lower potential economic costs.

In order to improve the botanical composition of native pasture in north-central region of Veracruz, Mexico, was evaluated two methods of establishment to incorporate the forage legume *Arachis pintoi* CIAT 17434.

5.1. Materials and methods

5.1.1. Location

Centre for Teaching, Research and Extension in Tropical Animal Husbandry of the Faculty of Veterinary Medicine, of the National University of Mexico (UNAM), located in the municipality of Tlapacoyan, Veracruz, Mexico, 20 ° 03 'north latitude and 97 ° 03' west longitude, 151 m. The climate is hot and humid on the type Af (m) (e), with an average temperature of 23.4 ° C and an average annual rainfall of 1980 mm. Soil characteristics are presented in the Table 12.

Properties	Soil depth (cm)			
	0-10	10-20	20-30	30-40
Texture (%)				
Sand	22.2	8.6	-	18.2
Clay	47.0	70.9	-	57.5
Silt	30.8	20.5	-	24.4
Chemicals				
pH	5.0	5.1	5.3	5.3
O.M. (%)	3.5	1.7	1.0	1.2
P (ppm)	5.0	6.4	4.4	2.0
S (ppm)	32.0	54.4	41.6	34.1
Ca (meq/100 g)	5.1	5.0	4.2	4.0
Mg	1.8	1.5	1.4	1.4
K	0.8	0.3	0.3	0.3
Al	0.2	0.1	0.1	0.1
CEC (meq/100 g)	7.1	6.8	5.7	5.5
Al saturation (%)	2.8	1.5	1.7	1.8

Table 12. Chemical and physical characterisics of the experimental soil. Veracruz, Mexico.

The study was conducted during the three seasons representative of this region: winter or "North" from November to February; drought: March to June, rain or summer: July to October. Weather conditions for the experimental period by time are presented in Figure 9.

Experimental design and treatments. We used a completely randomized design with factorial arrangement of 2 x 2: Conventional or minimum tillage, and fertilization or not, within each period and 12 observations (no repetitions) per treatment (T), resulting in:

T1 = Conventional tillage plus fertilizer

T2 = Conventional tillage without fertilizer

T3 = Minimum tillage plus fertilizer

T4 = Minimum tillage without fertilizer

T1 and T3 were: P (22), S (25), K (18), Mg (20), Ca (100), Zn (3), Cu (2) and B (1) kg / ha. Each period included a an experimental area, with dimensions of 50 m x 40 m (2000 m^2), divided into two parts along: one, conventional tillage; and another with minimum tillage. Then each part was subdivided again in width to the treatments with and without fertilization. Each treatment involved 12 observations (no repeats) within the corresponding area of 9 m^2 each for thorough preparation, and 18 m^2 for minimum tillage.

Land preparation. In T1 and T2, were allocated strips of 3 m x 20 m, alternating with native grass, where the vegetation was slashed with desvaradora, followed by 4 to 5 passes of harrow and plowed with a hoe. On the strips, the distance between rows and plants within them was 80 and 50 cm, respectively. The legume is seeded with a seed depth of 15 cm.

Minimum tillage. For T3 and T4, there was a land clearing with machete, were traced rows of 20 m long, spaced every two meters, and the rows were holes (seed points) every 50 cm, diameter and depth of 20 and 15 cm, respectively.

Planting dates were in Nov 29/1991, March 2/1992 and jul 15/1992, with plant material, placing 3-4 stems of 15 cm long, with only three or four leaves in the air. After 30 days, treatments were applied "with" and "without fertilization. These works were carried out in each season and in the corresponding area. Weed control was made with a hoe, in the first three months, for each treatment and time.

Variables. Data were collected at 4, 8, and 12 weeks post-seeding for number of plants, and 4, 8, 12 and 24 weeks for coverage. The useful area was 9 m^2 (T1 and T2) and 18 m^2 (T3 and T4). The first variable was the number of facilities within the useful area and in the second, the proportion was estimated visually apparent that the legume covered the area. The data were analyzed separately for each planting season, using ANOVA, and Tukey's test was used to compare means [9]. Regression coefficients were estimated to number of plants (linear) and coverage (exponential) to observe trends.

5.2. Results

5.2.1. Winter season

Number of plants. At this time, the average/treatment at 4, 8 and 12 weeks was 12.2, 15.3, 14.5 and 14.7 plants/9 m². A significant effect (P≤0.01) by fertilizer and ages was observed; by the contrary, the interaction week x treatment was not significant. The overall average was 14.1 plants/9 m² with a coefficient of variation of 20.0%. Treatments 2, 3 and 4 had better performance at 12 weeks.

Regarding the rate of appearance of plants, expressed this as the time in weeks to bring a new plant, was similar among treatments 2, 3 and 4 in winter and rainy seasons, while T1 needed more time to build a new plant (Table 13).

Planting season	T1	T2	T3	T4
	Weeks by treatment			
Winter	3,8	1.5	2.0	1.7
Dry	9.0	33.3	-	25.0
Rainy	1.0	0.6	0.5	1.0

Time in weeks to bring a new plant.

Table 13. Appearance rate* of *Arachis pintoi* plants on each planting season. Veracruz, Mexico.

Coverage. At 12 weeks, the best coverage was in T2 (P≤0.01) with 28%, while while T1 and T3 were similar, with 22.9% and 19.0% respectively. On the contrary, coverage at T4 was 18%. The overall average for this variable during the winter season was 22.1% with a coefficient of variation of 38%.

There were statistical differences between treatments (P≤0.05), exceeding 28% of T2 with coverage, while T1 and T3 were similar, with 22.9% and 19.0% respectively. The average for T4 was 18.2%. The overall average for this variable during the winter season was 22.1% with a coefficient of variation of 38%.

Figure 10A shows the increase in coverage during the establishment period, for each treatment at 4, 8, 12 and 24 weeks. There is a considerable increase for all treatments from week 12. The maximum coverage at 24 weeks is presented in conventional tillage treatments.

The rate of coverage of the ground, expressed as the average time in weeks for the plants to cover 10% of area, is presented in Table 14.

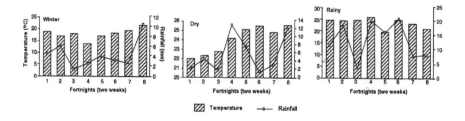

Figure 9. Temperature and rainfall on each one of the planting seasons of *Arachis pintoi*. Veracruz, Mexico.

Planting season	T1	T2	T3	T4
	Weeks by treatment			
Winter	2.9	2.6	3.8	4.2
Dry	4.7	4.2	3.5	2.2
Rainy	2.2	2.8	2.6	2.0

Table 14. Time in weeks to *Arachis pintoi* cover 10% of soil. Veracruz, Mexico.

5.2.2. Dry season

Number of plants. The averages for this variable were: 21.15, 19.75, 32.5 and 31.9 plants/9m^2 assessment considering each week. T1 and T2 were statistically equal, but different from T3 and T4 (P≤0.05).

In conventional tillage, were less plants than at minimum tillage treatments. Theere are not significance for age effect, neither its interaction with soil treatment Table 13.

5.2.3. Coverage

The best coverage (>25%) was at at 3 and 4 treatments (P≤0.05). Figure 10B shows the soil coverage at each evaluation frequency. An outsatndinh behaviour was observed for T1 after 8 weeks, achievinig 80% coverage to 24 weeks.

The age effect and its interaction with treatments were statistically significant. The shortest time to cover 10% of soil was during dry and rainy seasons at T4 (Table 14).

5.2.4. Rainy season

Number of plants. At this time, the largest number of plants/9 m^2 occurred at treatments 1 and 2 (23.0 plants), compared to T3 and T4 (17.3 and 14.3 plants, respectively). Was observed an increase of plants at 8 weeks, mainly in conventional tillage treatments. The shortest time or highest rate of occurrence of plants in T1 and T4 was 1.0 weeks in time for the emergence of a new plant.

5.2.5. Coverage

During the rainy season, soil coverage was similar among treatments (Figure 10C). The overall average was 34.5% with a coefficient of variation of 26.1%. For the rate of ground coverage, the lowest average time was observed in T4 with 2.0 weeks to cover 10% of the area (Table 14).

5.2.6. Number of plants in each season

The average of plants/m2 was largest during the rainy season (19.4 plants) followed by winter season (14.2 plants), and dry season (8,7 plants). The analysis of variance and regression coefficients for number of plants/season are shown in Table 15.

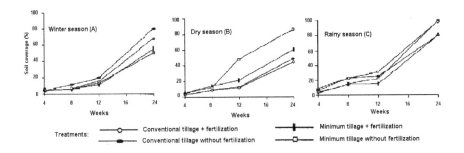

Treatments:
—○— Conventional tillage + fertilization   —■— Minimum tillage + fertilization
—●— Conventional tillage without fertilization   —□— Minimum tillage without fertilization

Figure 10. Percentages of soil coverage by *Arachis pintoi* planted in winter (A), dry (B) and rainy (C) seasons. Veracruz, Mexico.

Source of variation	df	Winter		Dry		Rainy	
		MS	P>F	MS	P>F	MS	P>F
Treatments (T)	3	64.85	0.0001	185.64	0.0001	681.34	0.0001
Weeks (lineal)	1	384.00	0.0001	2.34	0.5901	2860.16	0.0001
T x W (lineal)	3	11.58	0.2532	1.37	0.9161	68.80	0.3772
Error	136						
Coefficients		a	b	a	b	a	b
T1		10.08	0.26	6.14	0.11	14.80	1.03
T2		9.94	0.66	6.33	0.03	10.05	1.62
T3		10.49	0.50	11.07	-0.03	2.73	1.62
T4		10.11	0.57	10.30	0.04	6.50	0.98

Table 15. Analysis of variance and regression coefficient for number of plants by planting season of *Arachis pintoi*. Veracruz, Mexico.

5.2.7. Soil coverage and age of plants

The best percentages of soil coverage by A. *pintoi* occurred at 24 weeks, highlighting the rainy season planting date. Table 16 shows the analysis of variance and regression coefficients between soil coverag and age of plants on each season.

Source of variation	df	Winter		Dry		Rainy	
		MS	P>F	MS	P>F	MS	P>F
Treatments (T)	3	986.39	0.0001	4933.92	0.0001	696.84	0.0001
Weeks (lineal)	1	102416.09	0.0001	93319.28	0.0001	193155.43	0.0001
T x W (lineal)	3	1300.38	0.0001	2932.31	0.0001	958.32	0.0001
Error	184						
Coefficients		Winter season		Dry season		Rainy season	
T1		$Y=2.42(exp.^{0.1396x})$		$Y=2.42(exp.^{0.1095x})$		$Y=2.42(exp.^{0.1068x})$	
T2		$Y=2.42(exp.^{0.1405x})$		$Y=2.42(exp.^{0.1140x})$		$Y=2.42(exp.^{0.1064x})$	
T3		$Y=2.42(exp.^{0.1155x})$		$Y=2.42(exp.^{0.1095x})$		$Y=2.42(exp.^{0.1305x})$	
T4		$Y=2.42(exp.^{0.1133x})$		$Y=2.42(exp.^{0.1326x})$		$Y=2.42(exp.^{0.1472x})$	

Table 16. Analysis of variance and regression coefficient for soil coverage by planting season of Arachis pintoi. Veracruz, Mexico.

5.3. Discussion

Although the two ways to establish *Arachis pintoi* tested here are not the only ones, the results with conventional tillage are attractive, in the frequencies tested. In this regard, the method [46], using a planting implement, designed for them, allowed that two months after planting shown good development. Here, at 24 weeks, the ground cover in all treatments was above 80%, while the total coverage (100%) in treatments with conventional tillage was achieved approximately eight months post-planting.

Should be noted that the availability of plant material is an advantage in the evaluation of the species, as well as attempts to disseminate the same among low-income producers, because of the ease of material handling.

The null effect of fertilization on the establishment of *Arachis pintoi* found here, was also observed in Colombia [47], who applied seed and fertilizer pellets to a degraded pasture of *Brachiaria*. The fertilizers were the same as those applied here, except Zn, Cu and B, although in much smaller quantities. This lack of effect could be explained based on the relatively short period of observation (12 and 24 weeks for number of plants and coverage, respectively) per day, such as to indicate their presence nutrients to the crop, especially in the case of P, which is referred to their low mobility in soil.

A study in this same experimental field [48], reports that when *A. pintoi* was planted by seed coverage was achieved over 90% at 12 weeks of age. In Puerto Rico, accessions CIAT: 18744, 18747 and 18748 evaluated in an Oxisol had a high rate of spread at 16 weeks with 90% ground cover and low incidence of weeds [49]. This coverage is lower than that obtained in Colombia, were researchers established *Arachis pintoi* by vegetative associated with *B. dictyoneura*, and 20 weeks of age achieved a coverage of 40 to 45% [50].

During the experiment, the climate in CEIEGT, was very variable. This is because the region is in a climatic transition zone between the regions: coastal (subhumid) on the east, and Sierra Madre Oriental (wet), to the west, which creates a very unstable microclimate between and within years. Evidence of this was the precipitation that was 40% above the annual average and in the dry season, rainfall was 60% higher than that in the last 10 years. So, this season should be considered "atypical" and therefore the results may not be reliable. Unlike precipitation, the temperature is somewhat variable.

5.4. Conclusions

Based on the information presented, we conclude that: (1) establishing methods involving a conventional land preparation proved to be the best, both for number of plants to cover. (2) Although were not evaluated seasons, the best results were achieved in the r.ainy season (3) No effect was observed for either fertilization treatments. (4) whereas in the dry season rainfall was well above average, the results obtained at this time, must be taken with caution. (5) is suggested to evaluate Arachis pintoi in locations with different climate and soil, in addition to testing planting seasons. (6) must also consider the possibility of evaluating new methods of establishment, both plant material, as with sexual seed.

Author details

Braulio Valles-de la Mora*, Epigmenio Castillo-Gallegos, Jesús Jarillo-Rodríguez and Eliazar Ocaña-Zavaleta

*Address all correspondence to: braulio_36@yahoo.com.mx

Universidad Nacional Autónoma de México, Facultad de Medicina Veterinaria y Zootecnia, Centro de Enseñanza, Investigación y Extensión en Ganadería Tropical (CEIEGT), México

References

[1] Silva JE, Resck D.V.S., J. Corazza E., Vivaldi L. Carbon storage in clayey Oxisol cultivated pastures in the "Cerrado" region, Brazil. Agriculture, Ecosystems and Environment 2004;103(2): 357–363.

[2] SEMARNAT (Secretaría del Medio Ambiente y Recursos Naturales). Indicadores para la Evaluación del Desempeño Ambiental. Reporte 2000. Dirección General de Gestión e Información Ambiental. México, D.F.; 2000.

[3] Curti-Díaz SA., Loredo-Salazar X., Díaz-Zorrilla U., Sandoval R., Hernández J. Tecnología para producir limón Persa. Libro técnico No. 8. CIRGOC- INIFAP-SAGAR. Veracruz, México; 2000.

[4] Rincón AC., Orduz JO. Usos alternativos de *Arachis pintoi:* Ecotipos promisorios como cobertura de suelos en el cultivo de cítricos. Pasturas Tropicales 2004;26(2): 2-8.

[5] Clement CR., Defrank J. The use of ground covers during the establishment of heart-of-palm plantations in Hawaii. Horticulture 1998;33(5): 814-815.

[6] Johns GG. Effect of *Arachis pintoi* groundcover on performance of bananas in northern New-South-Wales. Australian Journal of Experimental Agriculture 1994;34 (8): 1197-1204.

[7] Dwyer GT. Pinto's peanut: a ground cover for orchards. Queensland Agricultural Journal 1989; (May-June): 153-154.

[8] SPSS (Statistical Package for the Social Sciences). Sigma Plot for Windows. Version 4.00. SPSS Inc. Chicago, Ill. 1997.

[9] SAS. "SAS/STAT User's Guide (Version 6)," 4th/Ed. SAS Institute Inc., Cary, N. C. 1989.

[10] Castillo GE. Improving a native pasture with the legume *Arachis pintoi* in the humid tropics of Mexico. PhD thesis. Wageningen University, The Netherlands; 2003.

[11] Pérez-Jiménez SC., Castillo E., Escalona MA., Valles B., Jarillo J. Evaluación de *Arachis pintoi* CIAT 17434 en una plantación de naranja var. Valencia. In: Argel P., Ramírez A. (eds.) Experiencias regionales con *Arachis pintoi* y planes futuros de investigación y promoción de la especie en México, Centroamérica y el Caribe. Cali, Colombia. Documento de trabajo no. 159. Centro Internacional de Agricultura Tropical; 1996.

[12] Granstedt R. Rodríguez A.M. Establecimiento de *Arachis pintoi* como cultivo de cobertura en plantaciones de banano. In: Argel P., Ramírez A. (eds.) Experiencias regionales con *Arachis pintoi* y planes futuros de investigación y promoción de la especie en México, Centroamérica y el Caribe. Cali, Colombia. Documento de trabajo no. 159. Centro Internacional de Agricultura Tropical; 1996.

[13] Staver C. *Arachis pintoi* como cobertura en el cultivo del café: Resultados de investigación y experiencias con productores en Nicaragua. In: Argel P., Ramírez A. (eds.) Experiencias regionales con *Arachis pintoi* y planes futuros de investigación y promoción de la especie en México, Centroamérica y el Caribe. Cali, Colombia. Documento de trabajo no. 159. Centro Internacional de Agricultura Tropical; 1996.

[14] Perin A, Guerra JGM, Texeira MG. Soil coverage and nutrient accumulation by pinto peanut. Pesquisa Agropecuaria Brasileira 2003;38(7): 791-796.

[15] Thomas RJ, Asakawa NM. Decomposition of leaf-litter from tropical forage grasses and legumes. Soil Biology and Biochemistry 1993;25(10): 1351-1361.

[16] Mannetje L.'t. Harry Stobbs Memorial Lecture, 1994 - Potential and prospects of legume-based pastures in the tropics. Tropical Grasslands 1997;31(2): 81-94.

[17] Argel PJ. Regional experience with forage Arachis in Central America and Mexico. In: Kerridge PC., Hardy B. (eds.) Biology and Agronomy of Forage Arachis. Cali, Colombia: CIAT; 1994. p134-143.

[18] Lascano CE. Nutritive value and animal production of forage Arachis. In: Kerridge PC., Hardy B. (eds.) Biology and Agronomy of Forage Arachis. Cali, Colombia: CIAT; 1994. p109-121.

[19] Hernández M, Argel PJ, Ibrahim MA, Mannetje L't. Pasture production, diet selection and liveweight gains of cattle grazing Brachiaria brizantha with or without Arachis pintoi at 2 stocking rates in the Atlantic zone of Costa Rica. Tropical Grasslands 1995;29(3): 134-141.

[20] Ibrahim MA, Mannetje, L't. Compatibility, persistence and productivity of grass-legume mixtures in the humid tropics of Costa Rica. 1. Dry matter yield, nitrogen yield and botanical composition. Tropical Grasslands 1998; 32(2): 96-104.

[21] Karbassi P, Garrard LA, West SH. Effect of low night temperature on growth and amylolytic activities of two species of Digitaria. Proceedings of the Soil and Crop Science Society of Florida 1970;30: 251-255.

[22] Ivory DA, Whiteman PC. Effect of temperature on growth of five subtropical grasses. I. Effect of day and night temperature on growth and morphological development. Australian Journal of Plant Physiology 1978a;5(2): 131-148.

[23] Ivory DA, Whiteman PC. Effect of temperature on growth of five subtropical grasses. II. Effect of low night temperature. Australian Journal of Plant Physiology 1978b;5(2): 149-157.

[24] Toledo JM. Plan de Investigación en Leguminosas Tropicales para el CIEEGT, Martínez de la Torre, Veracruz, México. Informe de Consultoría en Pastos Tropicales al Proyecto: Enseñanza y Extensión para la Producción de Leche y Carne en el Trópico. FAO/CIEEGT, FMVZ, UNAM, Martínez de la Torre, Veracruz, México (Circulación interna). 1986.

[25] Cook S, Ratcliff D. Effect of fertilizer, root and shoot competition on the growth of Siratro (Macroptilium atropurpureum) and Green Panic (Panicum maximum var. trichoglume). Australian Journal of Agricultural Research 1985;36(2): 233-245.

[26] Toledo JM., Schultze-Kraft R. Metodología para la evaluación agronómica de pastos tropicales. In: Toledo JM (ed.). Manual para la evaluación Agronómica. Cali, Colombia: Centro Internacional de Agricultura Tropical; 1982. p91-110.

[27] Steel RGD., Torrie JH. Principles and Procedures of Statistics: A Biometrical Approach. New York, USA: Mc Graw-Hill, Inc.; 1980.

[28] Núñez GLF. Evaluación biológica y económica en el establecimiento de *Arachis pintoi* como cobertera en cítricos con café. Tesis. Universidad Autónoma Agraria Antonio Narro, México; 1997.

[29] Chambliss GC., Williams MJ., Mullahey JJ. Savanna Stylo Production Guide. Florida Cooperative Extension Service. University of Florida, Gainesville, Florida, USA; 2000.

[30] Schulke B. Pasture establishment in the coastal Burnett, Queensland. 2000. http://www.dpi.qld.gov.au/beef/3313.html. (accessed 5 October 2004).

[31] Valles MB, Cadisch G, Castillo GE. Mineralización de nitrógeno en suelos de pasturas con *Arachis pintoi*. Técnica Pecuaria en México 2008;46(1): 91-105.

[32] Villarreal M., Vargas W. Establecimiento de *Arachis pintoi* y producción de material para multiplicación. In: Argel PJ, Ramírez A. (eds.) Experiencias Regionales con *Arachis pintoi* y Planes Futuros de Investigación y Promoción de la Especie en México, Centro América y El Caribe. Cali, Colombia. Centro Internacional de Agricultura Tropical; 1996.

[33] Argel PJ, Villarreal MC. Nuevo Mani Forrajero Perenne. Cultivar Porvenir CIAT 18744. San Jose, Costa Rica: MAG, IICA, CIAT; 1999.

[34] Stur WW, Horne PM. Developing forage technologies with smallholder farmers - how to grow, manage and use forages [Indonesian]. Monograph. Canberra, Australia. Australian Centre for International Agricultural Research (ACIAR); 2001.

[35] Pizarro EA, Rincón AC. Regional experience with forage *Arachis* in South America. In Kerridge PC., Hardy B. (eds.). Biology and Agronomy of Forage *Arachis*, Cali, Colombia: CIAT; 1994. p144-157.

[36] Rivas L, Holmann F. Adopción temprana de *Arachis pintoi* en el trópico humedo: El caso de los sistemas ganaderos de doble propósito en el Caquetá, Colombia. Pasturas Tropicales 1999;21(1): 2-17.

[37] Bosman HG, Castillo E, Valles B, De Lucía GR. Composición botánica y nodulación de leguminos en las pasturas nativas de la planicie costera del Golfo de México. Pasturas Tropicales 1990;12(1): 2-8.

[38] Valles B, Castillo E, Hernández T. Producción estacional de leguminosas forrajeras en Veracruz, México. Pasturas Tropicales1992;14(2): 32-36.

[39] Ayarza MA., Spain JM. Manejo del ambiente físico y químico en el establecimiento de pasturas mejoradas. In: Lascano C., Spain J. (eds.) Establecimiento y Renovación

de Pasturas. Sexta reunión del Comité Asesor de la RIEPT. Veracruz, México: CIAT; 1991. p189-208.

[40] Carvajal AJJ. Producción de semillas de cultivos de cobertura: *Pueraria phaseoloides*. Livestock Research for Rural Development 2009;21(39). http://www.lrrd.org/lrrd21/3/ carv21039.htm (Accessed 5 August 5 2012).

[41] Flores M. Bromatología Animal. Mexico, D.F.: LIMUSA; 1983.

[42] Garza TR, Portugal GA, Ballesteros WH. Evaluacion en pastoreo de asociaciones de zacates y leguminosas utilizando vaquillas de razas europeas en clima tropical. Tecnica Pecuaria en Mexico 1972;23(1): 7-11.

[43] Funes F. Effect of fire and grazing in the maintenance of tropical grasslands. Cuban Journal of Agricultural Science 1975; 9(3): 379-395.

[44] Sollenberger LE, Quesenberry KH, Moore JE. Forage quality responses of an *Aeschynomen-* limpograss association to grazing management. Agronomy Journal 1987;79(1): 83–88.

[45] Ruiz T, Febles G, Sistachs M, Bernal G, León J. Prácticas para el control de malezas durante el establecimiento de *Leucaena leucocephala* en Cuba. Revista Cubana de Ciencia Animal 1990;24(2): 241-246.

[46] Azakawa NM, Ramírez RCA. Metodología para la inoculación y siembra de *Arachis pintoi*. Pasturas Tropicales 1989;11(2): 24-26.

[47] Ogawa Y, Mitamura T, Spain JM, Perdomo C, Avila P. Introduction of legumes in *Brachiaria humidicola* pasture using macro-pellet. Japan Agricultural Research Quaterly 1990;23(3): 232-240.

[48] Hernández T, Valles B, Castillo E. Evaluación de gramíneas y leguminosas forrajeras en Veracruz, México. Pasturas Tropicales 1990;12(3): 29:33.

[49] Valencia E, Sotomayor-Ríos A, Torres C. Perennial peanut: establishment and adaptation on an Oxisol in Puerto Rico, United States. USDA, Agriculture Research Service;1992.

[50] Gil E, Alvarez E, Maldonado G. Distancia y distribución de siembra en el establecimiento de tres especies de *Brachiaria* asociadas con leguminosas. Pasturas Tropicales 1991;13(3): 11-14.

Chapter 3

Enhancing Soil Fertility for Cereal Crop Production Through Biological Practices and the Integration of Organic and In-Organic Fertilizers in Northern Savanna Zone of Ghana

James M. Kombiok, Samuel Saaka J. Buah and
Jean M. Sogbedji

Additional information is available at the end of the chapter

1. Introduction

In Ghana, it has been estimated that 60 % of the population makes their living from subsistence farming with an average of 27% living in extreme poverty (MoFA, 2002). This is because the most dominant economic activity of the area is agriculture and once agriculture is not well developed, one of the effects is poverty. As observed by many, one of the characteristics of underdeveloped agriculture is the dominance of subsistence farming in these regions (MoFA, 2002, RELC 2004). The slow economic growth and high poverty level prevailing in Northern Ghana (Upper East, Upper West and the Brong-Ahafo regions) is therefore directly linked to the underdeveloped agricultural sector of the area.

The most affected area in the country is Northern Ghana as it is estimated that up to 80% of the population in this part of the country is poor (Ekekpi and Kombiok, 2008). The many agricultural interventions to transform the small scale farming system and reduce poverty in northern Ghana have largely failed due to several problems such erratic rainfall and poor soils. Further analysis of the northern Ghana agricultural sector problems indicates that poor soils result in low crop yields which are negatively affecting the development of Agriculture (RELC, 2005).

It is therefore not surprising that low soil fertility has always been mentioned by farmers as one of the constraints affecting cereal production in Northern Ghana (RELC, 2005). This is confirmed by the fact that low grain yields of cereals attributed to poor soils for the past decade has been ranked first among the constraints collated from all the districts of northern Ghana at the various regional planning sessions.

The low soil fertility in this part of the country is therefore blamed on the bush fires which usually occur annually during the dry season commencing from October to April the following year (SARI, 1995). This situation renders the soil bare exposing it to both wind and water erosion in the dry and rainy seasons respectively thereby depleting the macro-nutrients such as Nitrogen, Phosphorus and Potassium (NPK) and organic matter from the soil.

Initially, farmers used to replenish the soil with its nutrients by practicing shifting cultivation or land rotation. However, with the increase in population which has put pressure on land use, this practice is not being sustained and this therefore calls for other measures to maintain soil fertility for sustainable crop production in the savanna zone of Ghana.

The purpose of this chapter is therefore to expose to Agricultural science teachers/trainers, scientists and farmers:

- to the available soil fertility enhancing practices applicable in the Savanna zone of Northern Ghana.

- To discuss and recommend for adoption the most proven practices involving organic and in-organic materials either by applying each individually or the combination of both in the management of soil fertility for crop production in the Savanna zone of Ghana.

2. Materials and methods

Materials used were the various works done in the area of soil fertility management in the Savanna regions within the sub-Saharan Africa. These are published books, journal papers, annual reports and technical reports. It also included works done by the author, students dissertation supervised by the author and personal experiences gathered. Success stories from other interventions by the Government and Non-governmental Organizations implemented in the form of projects to raise soil fertility status in the zone were also considered. All were reviewed, discussed and conclusions drawn from the results of these various practices.

The various interventions being practiced within the Sub-region to enhance or maintain the soil fertility status include the manipulating of the crops planted (cropping systems) in the area. Others are land rotation, conservation agriculture and the application of different types of soil fertility enhancing materials and the integration of some or all of these as a single treatment. The difficulties associated with adoption of these practices will also be outlined. Some of these are:

3. Land rotation or shifting cultivation

It is a crop production system based on rotation of cultivated period on a given piece of land. The cultivated period is always shorter than the fallow period because the system is characterized by the use of very little or no external soil fertility improving inputs. The soil fertility is therefore recovered by a natural process which is often very slow. The length of fallow period is determined by land availability but can last between 10 and 20 years after which the vegetation is cut back to allow another cycle of farming activities(EPA, 2011). This is no longer practiced because of the scarcity of arable land as a result of high growth in population of the country.

4. Cover cropping and improved short-season fallow with leguminous cover crop

The practice of planting certain crops to cover the cultivated area of fallow land thereby providing protection for the purpose of reducing erosion by rain drop splash and surface runoff and weed growth is referred to as cover cropping. Where necessary, cover crops are cut down or killed by weedicides so that the seasonal crops can be planted in the mulch.

Improved short fallow is the planting of leguminous cover crops consciously with the objective of protecting the soil surface and fixing nitrogen as part of the crop fallow. The system is practiced where land is limited and farmers rely on little or no external soil enhancing materials to improve the soil fertility status. Examples of cover crops: *Mucuna pruriens, Dolicus lablab Canavalia ensiformis.* Sometimes edible cover crops such as the creeping types of cowpea are used.

Improved fallow systems using e.g. *Mucuna puriens* are promoted by different projects in Ghana. So far, the system seems to be adopted only by some farmers in certain areas (Quansah et al., 1998). Benefits observed by the farmers may vary and include increased soil moisture, weed suppression and residual yield effects on maize. Leguminous fallows have been used in northern Ghana to accumulate N from biological N fixation (BNF), smother weeds, and improve soil physical properties (Fosu et al., 2004). Leguminous cover crop systems apparently were more extensively tested than tree fallow systems in the country. Demand for arable land in many parts of the country has increased in recent years in response to increasing human population. This situation is gradually moving the emphasis from resting fallow to continuous and intensive cropping. However, some farmers in southern Ghana can still fallow their lands up to two years or two seasons. Leguminous cover crops such as *Callopogonium, Dolichos, Mucuna* and *Cannavalia* species are the main cover crops used during this short fallow. In the north, farm lands to undergo short term fallow of about two years were planted to either *Mucuna* or *Callopogonium* and left for these number of years. In the third year the residue was either ploughed in or if it was dry, the crop was planted directly into the mulch. This system was found to increase maize yields in both on-station and on-farm trials (Kombiok et al, 1995) as seen in Table 1. The highest mean yield of maize was obtained when residue was left but maize planted in it and this could be due to the micro-climate created by the residue on the surface and the decay of some of the residue to make N available to the crop.

Treatment	0 N kg/ha	40 N kg/ha	Mean
1-Residue not removed /No till	4110	4630	4370
2-Residue removed/No till	3043	3825	3434
3-Residue worked in by hoeing	3542	5176	4359
4-Residue removed/hoeing	3249	4249	3749
Mean	3486	4470	3978

Source: Kombiok et al 1995

Table 1. Maize Yield (kg/ha) under 4 different Management Practices after 2 years of *Callopogonium* fallow and three rates of nitrogen fertilizer at Nyankpala

In the wetter southern parts of the country, *Cannavalia or Mucuna* is planted as a minor-season fallow, from August to March. In the next major season beginning in April, farmers plant their crops (maize, yam, cassava or any other crop) through the mulch without burning. Weed control in the crops is by the hoe in northern Ghana and cutlass in southern Ghana when necessary. Where rainfall during the major season is not reliable, farmers plant the cover crop in April, and plant the food crop in the minor season (August–September). A synthesis of results of trials of a *Mucuna* fallow system by Carsky et al. (2001) suggested that in simultaneous intercropping systems, the yield of maize associated with *Mucuna* is decreased dramatically as the Mucuna smothers the maize. However, maize yield reduction from relay intercropping of Mucuna at 40 to 50 days after maize planting is only about 5%.

Similarly, in a 2-yr study on a typical plinthic Planleustalf in the savanna zone of Ghana, Kombiok and Clottey (2003) found that maize grain yields obtained after two years of interplanting *Mucuna* was highest at 6 weeks after planting followed by 8 weeks after planting and the least was obtained from 10 weeks after planting (Table 2). It was also found that N was highest in the plot with *Mucuna* planted 6 and 8 weeks after maize compared to 10 weeks and the bush fallow treatments due to the amount of biomass produced by the cover crop. They concluded that the highest yields of maize from relay cropping of *Mucuna* at 6 weeks after maize was due to the beneficial effects of the decay of the higher *Mucuna* biomass produced in that treatment in the previous years. It is clear that if *Mucuna* does not accumulate substantial biomass, then it will not accumulate sufficient N as well as suppress weeds. They however recommended the pruning of Mucuna as a management strategy to ensure it does not smother the maize crop in association with the cover crop.

Treatments	Maize grain yield (kg ha⁻¹)[1]		
	1996	1997	1998
Control	1800 a	1180 b	1050 d
6 WAP	1500 b	1620 a	1850 a
8 WAP	1130 c	1110 b	1650 b
10 WAP	1100 c	1530 a	1250 c

[1]For a factor. means followed by a similar letter in a column are not significant at 5% level of significance

[2]WAP. Weeks after planting

Source: Kombiok and Clottey 2003

Table 2. Maize grain yield as affected by time of interplanting mucuna in maize and after two years of continuous mucuna

In some cases where weed infestation is high, farmers still have to do one weeding. After establishment, the *Mucuna* survives the short dry period during July–August and later forms a thick biomass that peaks in mid-November. This biomass canopy then covers the soil until it starts to decompose in the dry season in December– January. *Mucuna* seeds are then harvested and stored for later use. Unlike the late-maturing *Mucuna* variety, planting the medium- to early-maturing mottled variety is done at the onset of the minor season (Loos et al. 2001). Clearing and at least one initial weeding may be necessary for successful establishment. Once established, the biomass covers the soil surface and dies back naturally with the onset of the dry season. This means that no other food crop can be planted on the land that has an improved *Mucuna* fallow. After the improved fallow, farmers plant any other crop such as maize, yam, cassava or plantain. Incorporating *Mucuna* into the local cropping systems by intercropping *Mucuna* with plantain during the first season can be very economical. Such a system reduces the overall demand for labour, as it requires only spot weeding of *Mucuna* vines at certain intervals.

In places where rice cultivation is significant, farmers have developed and adapted *Mucuna*–rice rotations. Here, farmers cultivate rice in the major season and follow it with *Mucuna* in the minor season to suppress weeds and improve the fertility of the soil. *Canavalia ensiformis* did not attract the same attention as *Mucuna* because it was less vigorous in growth and did not suppress weeds as well as *Mucuna*. The less aggressive nature of *Canavalia* made it an ideal cover crop to use in mixed-cropping systems.

5. Cropping systems

A cropping system may be defined as a community of plants which is managed by a farm unit to achieve various human goals (FAO, 1995). In this particular case the cropping system is to achieve an enhanced soil fertility status for increased crop production.

5.1. Multiple cropping

It is the growing of different arable crops and /or other crops on a given piece of land at the same time. The aim is to increase the productivity from the land while providing protection of the soil from erosion. Growing more than one crop at the same time also cushions the farmer against total crop failure as adverse growing conditions might not affect the different crops equally-*sequential cropping* (growing two or more crops on the same piece of land within the same year or season but planting one after harvesting the other) or *intercropping* which is the growing of two or more crops on the same piece of land at the same time (Abalu, 1977). The existence of multiple cropping especially intercropping system involving mostly cereals and legumes among the small scale farmers of West Africa has long since been identified (Norman, 1975) and studied by many workers including Andrews and Kassam (1976), Fisher (1979) and Willey (1979).

Some of the reasons advanced for the persistence of this system of cropping have been precautions against uncertainty and instability of income and unstable soil fertility maintenance (Abalu, 1977). In most of the intercropping trials implemented in the sub-region the results of the crop yields showed that there have been agronomic advantages in the practice since the Land Equivalent ration (LER) is always more than one (1). In addition to the agronomic advantage in terms of yield associated with intercropping systems, a substantial amount of N is also fixed by the leguminous component of the system (Table 3).

Cropping System	Maize grain		Cowpea grain		%N fixed	
	2000	2001	2000	2001	2000	2001
Sole	2734	2400	1401	1153	40.47	62.42
Inter	1069	1731	954	544	30.20	34.74
Intra	-	1938	-	473	-	28.72
SE	124	111	52	28	6.08	10.89
LSD (0.05)	398	355	166	91	NS	NS

Source: Kombiok et al 2005

Table 3. Grain yields (kg ha^{-1}) of maize and cowpea and percent N fixed as affected cropping Systems 2000 and 2001

It was found that more than 50% N was fixed by the component cowpea in maize cowpea mixture. This is very beneficial to the farmers since the cereal crop component of the system will benefit from this N fixed if the legume matures earlier than the cereal. Secondly, there will also be a residual N left in the soil for use by any subsequent crop grown on the same piece of land in the next cropping season.

5.2. Crop rotation and intercropping

Cereal production in Ghana, especially northern Ghana is limited by low levels of nitrogen in the soil. Strategies such as intercropping/mixed cropping and crop rotations involving cereals and legumes have been adopted to raise crop yields as they fix substantial amounts of atmospheric N, can provide large amounts of N-rich biomass. Legumes grown as a food crop or live mulch (cover crop) can be successfully rotated with a crop which produces high biomass or intercropped with tree species (e.g. alley cropping) in order to provide N, enhance organic matter content and agroforestry. The amount of N returned from legume rotations depends on whether the legume is harvested for seed, used for forage, or incorporated as a green manure.

Crop rotation entails the growing of different crops in a well defined sequence on the same piece of land- Changing the type of crops grown in the field each season or year. Eg; a field could be planted to maize in the major season as in the south of Ghana and after harvesting, the same field is planted to cowpea in the minor season of the year. In the savanna region this will be done yearly since there is only one rainy season/cropping season in a year.

Crop rotation forms a central pillar of CA, and many approaches highlight the use of cereal–legume rotations. Rotations allow crops with different rooting patterns to use the soil sequentially, reduce pests and diseases harmful to crops and sustain the productivity of the cropping system. The most widely grown legumes in the farming systems of Ghana are the grain legumes; groundnut, cowpea and soybean. These crops have the advantage over other legumes in that they provide a direct economic yield for food or for sale. Yet unless there is a ready market for the grain, farmers tend to grow grain legumes on only a small proportion of their land, and certainly not sufficient to provide a rotation across the farm. Analyses in northern Ghana, where farmers indicated their normal rotation is cereal/legume, showed that the actual area sown to the legume was often less than 30% of the farm area. Further investigation indicated that crop rotations tended to be practiced more on the fertile 'home-fields' than on the poorer outfields.

The yield response of cereal crop following a legume can be substantial. In Ghana, the grain yield of sorghum crop following groundnut averaged 30-40% higher than the yield of continuous sorghum (Schmidt and Frey, 1992; Buah, 2004). Horst and Härdter (1994) showed large maize yields in northern Ghana following cowpea. In all the cases, crop residue was not removed from the field after harvest. Nonetheless, crop residues are often removed from the field at harvest so they do not provide the mulch cover wanted for CA. Various field experiments have shown that crop rotation of maize with various legumes was beneficial for maize production and that maize following groundnut often had the greatest yields when compared with maize following other legumes (Härdter, 1989; Horst and Härdter, 1994; Schmidt and Frey, 1992). Cotton-maize rotation is the most common rotation system in the northern part of Ghana. Cotton, even though not a legume, its production is accompanied by the application of inputs such as fertilizers and chemicals. Maize is therefore planted after cotton to take advantage of the residual fertilizers applied in the previous year. Farmers have reported increases in maize yields in the north by several tons per hectare as a result of cultivating maize after cotton in a rotation system. In southern Ghana where there are two cropping seasons, maize is planted in the major season (April-June) and an edible legume such as cowpea or a cover crop (*Mucuna*) is planted in the minor season.

A legume as a candidate crop in intercrop systems is again being encouraged because of the same reason as above. In the northern part of Ghana where the soils are low in both organic matter and essential nutrients, farmers intercrop cereals with legumes. The most common intercropping systems in this area are maize/cowpea, millet/cowpea, and maize/soybean. In some cases, both in the north and south of Ghana cover crops such as *Mucuna or Callopogonium* is planted in maize at the latter part of its growth cycle (6 weeks after planting maize). In southern Ghana, maize is harvested earlier and the cover crop (*Mucuna*) is left to grow into the minor season (August to March) until the next major season (April to August). In the northern savanna zone however, the cover crop dries when the rains end in October and the residue forms mulch protecting the soil. The incorporation of the residue in the soil after two years of cropping increased both the soil nitrogen and maize grain yield significantly (Kombiok Clottey, 2003).

One approach that has proved to be inherently attractive to farmers and is standard practice in most parts of northern Ghana is intercropping maize or sorghum with the grain legume cowpea or groundnut. If cowpea is sown between maize rows, the plant population and yield of maize can be maintained, whilst reaping the advantage of yield from the cowpea harvest. There is high labour requirement in the practice of intercropping because more than one crop is being planted at a time. So, labour is required for planting the component crops and for the careful control of weeds in the system. Insecticides are needed for the control of insect pests on the legume component either being rotated or intercropped.

Pigeon pea is an ideal legume for intercropping with cereals. Its slow initial growth affords little competition with the cereal for light or water, and it continues growing into the dry season after the maize crop has been harvested. The leaves that fall from pigeon pea before harvest provide a mulch and can add as much as 90 kg N/ha to the soil that then mineralizes relatively slowly during the subsequent season, releasing N for the next maize crop (Adu-Gyamfi et al., 2007). Thus a substantial rotational benefit, although not a perfect soil cover, can be achieved for the next season.

6. Conservation agriculture technologies for soil fertility management in Ghana

Conservation Agriculture (CA) is described as a set of practices or procedures carried out that ensure higher agricultural productivity and profitability whilst improving soil health and environment. It is known to be hinged on three basic principles which are (i) Little or no disturbances of the soil, (ii) The soil should have a cover all year round and (iii) the crops should be in rotation from season to the other or in intercropping situations.

Conservation Agriculture was introduced into Ghana in the early 1970s, mostly through donor funded Agricultural projects. Even though results in terms of crop yields from the various on-farm experiments have been found to be higher than the yields from the traditional slash and burn method of farming, it has not been easy to convince farmers to adopt the practice holistically. Generally the adoption of CA by farmers in Ghana is therefore low and those who are said to have adopted CA may either be practicing one or two of the principles of CA such as no-till, no-till with intercropping but not all the three principles of the practice. Comparatively, climatic and weather conditions in the southern part of the country favour the adoption of some of these principles. For example, the rainfall system in the south is bi-modal with only a dry period of less than three months. This allows the growth and development of vegetation all year round and therefore not prone to bush fires. The decay of these vegetative matter when killed by weedicide, will go a long way to enrich the soil with its nutrients.

On the other hand, the northern part of Ghana has only one rainy season which commences in late May and ends in early November with a dry period of about five months which is characterized by the Hamattan winds. During the dry season, the vegetative matter is dried up and therefore prone to bush fires. The occurrence of bush fires either accidentally or intentionally, clears up all the dry vegetative cover exposing the soil to the Harmattan winds in the dry season and the running water during the rainy season which robs the soil of its nutrients. It is therefore not surprising to know that farmers in the northern part of the country consider any technology that conserves soil and water such as soil bunding as a CA technology. Table 4 shows the results of a survey that was carried out to identify technologies related to CA practiced by farmers in the northern part of Ghana. Comparatively, among the districts covered, East Mamprusi recorded a higher percentage of farmers practicing some of these technologies than Lawra or Bawku which according to Ekekpi and Kombiok (2008) could be indicative of better extension services in that district.

Technology	East Mamprusi (%)	Lawra (%)	Bawku (%)
Contour bunding	19	47	19
Crop rotation/intercropping	60	3	4
Agro-forestry	3	16	3
Manure/refuse application	22	16	19
Minimum tillage	3	0	0
Crop residue management	31	3	2
Composting/application	50	63	50
Organic farming	0	0	0
Animal traction	4	22	20
Rotation kraaling	3	0	0
Bush fallow	3	0	0

Number of respondents = 32

Source: Ekekpi and Kombiok, 2008

Table 4. Percentage of farmer respondents on CA technologies in the savanna zone

6.1. Direct seeding

Direct seeding of crops is carried out without tilling the land in most parts of Ghana. In the southern part of the country (forest zone) where vegetation exists all year round, the vegetation is slashed using a cutlass and the residue instead of burning, is left as mulch on the farm. However, in the north, the crops (maize, sorghum and millet) are planted directly on the bare soil since all vegetative matter would have been burnt during the dry season. In both cases, the crops are planted using a dibbling stick or cutlass to create holes either on the bare soil (North) or inside the mulch as in the case of the South. This is advantageous in the South since the farmer will benefit from the mulch as it will conserve soil water and eventually decay with time to add nutrients to the soil within the season. Under such a situation, the physical and biological properties of the soil are also expected to improve after the mulch decomposes. In the north however, weeding should be carried out within three weeks after planting after which fertilizer or manure would be applied to the farm since the soil is devoid of vegetative cover at planting. This is to be sure that the crops are supplied with enough nutrients and to avoid heavy weed infestation on the farm which can reduce crop yields.

6.2. Minimum or reduced tillage

Minimum tillage is the reduction in the number of times the soil is being tilled as in conventional tillage method (ploughing/harrow/ridge) before and after the crop is planted. In Ghana, most farmers have adopted the use of weedicide to reduce tillage for land preparation for crop production because of its additional benefits of reducing labour cost. Other benefits of minimum tillage include the reduction in energy costs and it enhances the organic matter content of the soil while conserving the soil.

The vegetation is either slashed or sprayed with weedicides followed by either burning the dead vegetation as done in the forest and transitional zones of Ghana before crops are planted. This implies the land is not tilled before planting. However, weeds in this system are controlled by the use of cutlasses in the south while in northern Ghana, this done by hand hoe or the use of bullocks thus reducing the number of times the soil is tilled. The use of hoe as practiced in the north in weed control helps to bury the young weeds that have just emerged after planting which easily decay and return nutrients in to the soil for crop use. The burning of the vegetation before planting the crop by farmers in the south has been discouraged since the full benefits of mulch which include improved moisture infiltration to reduce soil erosion will not be realized in such a situation (Wagger and Denton, 1992).

It is common to see farmers in northern Ghana planting annual crops such as maize, millet and sorghum on the old ridges constructed in the previous year. In the southern part of Ghana, it is the use of *glyphosate* (a total weed killer) at the recommended rate of 3l/ha which can be increased if noxious weeds such as spear grass (*Imperata cylindrical*) is present. It is advisable for farmers to delay planting of their crops for at least one week after the application of *glyphosate* to allow the breakdown of the chemical and to identify the portions not well treated. The dead weeds are either buried or left on the surface of the soil as done in the south. In both cases however, the number of times the land is physically tilled is reduced since tillage activity before planting the crop is avoided and this helps to maintain the structure of the fragile soil of the area.

In some cases, the application of pre-emergent weedicides such as Atrazine for maize production can also help to delay or avoid the use of hand hoe to remove weeds after planting of the crop. Comparing the north and south, not much fertilizer is applied in the south probably because the fertility of the soil is always improved after the mulch decomposes in the subsequent seasons. For high yields of crops in the north, application of higher rates of fertilizers is required.

6.3. Alley cropping with cover crops

Alley cropping is not widely practiced in Ghana but this is found in few places in the southern part of Ghana. It is similar to agro-forestry systems where fast growing shrubs or trees such as pigeon pea is planted in alleys while cover crops such as *Mucuna* or *Calapogonium* spp are planted to protect the soil from erosion and for weed control. Afetr harvesting the pigeon pea in the alleys, the biomass is harvested and used as mulch on the cover crops where maize is planted directly in the mulch. In this system, nitrogen is fixed in the soil from atmosphere by both the pigeon pea and the cover crops. Also, the decay of the biomass from the pigeon pea goes to enhance the N status of the soil which goes to improve the yield of the maize. Soil water is conserved and weeds are controlled effectively under this system.

6.4. Strip cropping

The planting of alternating strips of several crops aligned on the contour in the field is known as strip cropping. It is an effective conservation measure on slope between 5 and 10%. In this case, erosion is largely limited to the row crop strip and soil removed from these is trapped in the next strip down slope which is generally planted to close growing crops.

Strip cropping involving pigeon pea has many additional advantages especially in a mixed farming situation. In northern Ghana, almost every farm family raises livestock (goats, sheep and cattle) as well as poultry in addition to crop production (SARI, 1995). It is therefore common to find strips of pigeon pea on most fields where the grain is harvested and cooked on the farm as lunch for the family. Studies have shown that the biomass of pigeon pea can be pruned over three times within a year and shade-dried to feed livestock during the dry season. Among the three pruning heights of pigeon pea at 30, 60 and 90 cm in the trial, it was found that pruning at 90 cm height for livestock, the pigeon pea would still be able to produce seed at the end of the year which would not be significantly ($p<0.05$) different from the plant that was not pruned (Table 5). With the exception of the pigeon pea pruned at 30 cm, which produced significant highest amount of litter, the quantity of litter produced by those pruned at 60 cm and 90 cm were similar.

Pruning height of Pigeon pea (Fallow)	Dry Matter (litter) tons/ha		Grain yield of maize Tons/ha		Seed yield of pigeon pea (tons/ha)
	1997	1998	1998	1999	1998
Pigeon pea (no pruning)	2.13	4.27	3.32	1.16	1.11
Pigeon pea (30 cm)	0.37	0.73	3.42	0.69	0.06
Pigeon pea (60 cm)	0.87	1.77	2.80	0.93	0.36
Pigeon pea (90 cm)	1.67	1.97	2.53	0.93	0.62
Source: Agyare et al, 2002					

Table 5. Effect of pruning on leaf litter production, grain yield of pigeon pea and maize yield after two years of fallow in the northern Savanna zone of Ghana.

It was found that maize yield after two years of pigeon pea fallow was highest in 1998 at 30 cm pruning height which was followed by the pigeon pea not pruned at all in the trial. But maize yields from plots with pigeon pea pruned at 60 and 90 cm heights were similar in value but significantly lower than the yield obtained from the plots with pigeon pea pruned at 30 cm. The high maize yields at no pruning and the pruning at 30 cm height were attributed to the higher quantity of litter fall from the pigeon pea.

It was then concluded that biomass obtained from the pruning of pigeon pea up to 60 cm will be able to provide sufficient fodder that may be used to supplement livestock feeding in the dry season (Agyare et. al., 2002). This situation would not be sacrificing much in terms of soil fertility status, pigeon pea grain yield and yield of subsequent maize crop. This option makes pigeon pea a valuable leguminous shrub for short season fallow for the mixed farmer.

6.5. Agroforestry

Agroforestry involves the integration of trees/shrubs and sometimes animal husbandry in the farming system. It combines annual crops with herbaceous perennials or trees on the same units.

Both exotic and local tree species were screened for Agro-forestry purposes in SARI as from 1985 (Table 6). The results showed that *Gliricidia* and *Leucaena* which are both exotic tree species are better trees for soil fertility restoration than the local tree species like *Parkia*. Both the *Leucaena* and *Gliricidia* produced enough biomass much earlier for incorporation than the rest of the tree species, It was also found that the incorporation of pruned biomass from the tree species was responsible for the increase in soil nitrogen. This therefore suggest that the faster the growth and development of the tree species to produce biomass for incorporation, the better the tree for agro-forestry system.

N fert. rate (kg/ha)	Acacia	Leucaena	Parkia	Gliricidia	Vilellaria	Check/ control	Mean
0	1397	1397	1533	2240	1227	1987	1625
40	2250	1960	1937	2623	1887	2860	2253
80	2203	2643	2333	2630	1833	3180	2471
Mean	1950	1990	1935	2498	1649	2676	2116

Source: SARI, 1985

Table 6. Maize grain yield (kg/ha) under agroforestry system at SARI.

7. Application of fertilizers

The most common of the materials used as soil fertility enhancing substances however, are the organic and in- organic fertilizers. The recommended rates of in-organic fertilizers for the production of cereals especially maize in Ghana are the basal application of compound fertilizer made up of 15 % each of Nitrogen, Phosphorus and Potassium (NPK) at planting or two weeks after planting of 2 fifty kilograms (50 kg) bags per acre. This is followed by the application of either sulphate of Ammonia or urea at 1 fifty kilogram bag (50kg bag) or twenty-five kilogram bag (25 kg bag) per acre respectively just before the tasseling of maize. However, the acquisition of these materials whether the organic or in-organic fertilizers by farmers have also been faced with a lot of challenges.

In the first place, most of the small scale farmers are poor and cannot afford the recommended rates of the in-organic fertilizers to increase their crop yields. Most often, they just purchase the quantities that they can afford which are far below the recommended rates for the crops and therefore those quantities are unable to increase their yields. As a result, their crop yields still remain low and that explains why they remain poor.

Secondly, even though almost every farm family in northern Ghana possesses few livestock such as cattle, sheep, goats and or poultry, the dung (manure) they produce is highly inadequate to fertilize an area of one acre. Most of these categories of livestock are also on free range thereby making the gathering of their dung very difficult. In addition, some of the farms are very far from their homes so carting these bulky materials to their farmlands posses another challenge.

The above situation where farmers cannot afford recommended rates of in-organic fertilizers because they are poor and they also do not have enough animal dung to fertilize their crops call for the combination of both.

Maize grain yields (tons/ha)			
Treatments	Bunkpurugu	Walewale	Karaga
Tillage			
Bullock	0.98	1.21	0.41
Manual	1.06	0.85	0.08
LSD (0.05)	**0.29**	**0.58**	**0.38**
Fertilizers			
NPK	1.12	1.46	0.30
Manure (6t/ha)	0.70	0.99	0.16
1/2 rates (manure &NPK)	1.00	1.09	0.37
FP (No NPK/No manure)	0.54	0.60	0.24
LSD (0.05)	**0.27**	**0.31**	**0.33**

Source: CSIR-SARI 2007 Annual report

Table 7. Effect of tillage and fertilizer on maize grain yields at Bunkpurugu, walewale and Karaga

Studies have been conducted on the effect of tillage and fertilizers on the yield of maize for three consecutive years in three communities of the northern part of Ghana by the Savanna Agricultural Research Institute (SARI). Results confirmed that the application of the combination of half the rates of the organic and recommended in-organic fertilizers was as good as the application of the recommended in-organic fertilizers (Table 7) This suggests that, farmers with limited number of livestock or poultry can always supplement the manure they generate from these animals with half the rates of the recommended in-organic fertilizers to obtain high crop yields. The results however showed that there was no significant difference in yield of maize between the bullock and manual tillage systems indicating that any of the tillage systems will give similar maize yields.

Similarly, it has been found that the household waste generated and deposited outside the houses for several years are as rich in nutrients as the animal manure. Kombiok et al 1995 compared the yields of maize fertilized by animal dung and household waste in four communities of the East Mamprusi District of the Northern Region (Table.8).

Community	Yield ton/ha		
	R D(refused dump)	A M (Animal dung)	No fertilizers
Yaroyili	3.2	3.4	0.60
Bowku	2.7	2.8	1.33
Boayini	2.4	3.2	1.06
Tangbini	2.3	1.3	1.00
Average	2.65	2.68	0.99

Source: Kombiok et al 1995

Table 8. Effect of refuse and animal dung on the yield of maize at 4 sites in West Mamprusi District

The results showed that in some of the communities, the yields of maize under the animal manure and the household refuse were similar suggesting that both materials could contain similar quantity of nutrients. The use of these as soil fertility enhancing materials will not only increase crop yields but will also help to improve the sanitation status of these communities since all these heaps would be carted to the farms.

8. Effect of some soil fertility enhancing interventions on soil nutrients (NPK)

Table 9 shows the nutrient (NPK) values before and after some soil fertility enhancing interventions initiated by scientists within the Savanna zone of Ghana. The initial values of N in particular show that the highest was 0.049% and the lowest was 0.022% within this zone. These values are percent total nitrogen and not available N which means that not all these will even be available to the plant. The low N content of these soils therefore explains why yields of cereal crops are very low and in some cases no yield is obtained if no soil fertility enhancing material is applied to the soil. Results from omission trials carried out in Nyankpala for three consecutive years (2002-2005) showed that among the three major plant nutrients, nitrogen was the most limiting element for maize production (SARI, 2005)

Type/period of intervention	N%		P (ppm)		K (ppm)	
	Initial soil N	N After intervention	Initial Soil P	P After intervention	Initial Soil K	K After intervention
Intercropping *Mucuna* in maize (6WAP)	0.024	0.043	15.59	11.96	49.90	45.65
Effect of pigeon pea pruned at 30cm	0.028	0.095	14.89	19.25	46.36	52.85
2 years of *callpogonium*. fallow	0.049	0.062	13.60	23.90	39.58	42.62
The application of house hold refuse	0.022	0.085	14.83	20.08	45.60	48.45
The application of manure (cow dung)	0.028	0.092	16.58	22.65	40.01	44.82
Improved fallow *Mucuna* (1 year)	0.026	0.088	18.65	20.80	42.60	46.25

Table 9. Effect of different soil fertility enhancing interventions on soil NPK values within the Savanna zone of Ghana

With a minimum of two years of the various interventions however, there were increases in the nutrient (NPK) values which is indicative of the positive influence of these interventions on these elements in the soil. In most of the studies, the N values after the interventions were significantly ($p < 0.05$) higher than the initial N values but for P and K, there were no significant differences between the initial and after two years of intervention. The significant increases of percent N in the soil as a result of the various interventions also show how limited nitrogen is in the savanna soils.

9. Average farmer yields of some selected cereal crops Ghana

The average cereal yields of farmers in Ghana are very low. There may be many causes to the low crop yields obtained by farmers in Ghana. These include the use of local crop varieties which are low yielding; poor management of the crop on the field (late weed removal, inadequate plant population, late harvesting) but paramount among these is low soil fertility. This is because the same variety used by farmers without adequate supply of plant nutrients have been found to yield lower than the same variety properly managed by research scientists including the provision of adequate quantities of nutrients especially nitrogen.

Table 10 shows some of the average yields of cereal crops by farmers in Ghana as against the yields obtained from properly managed fields with adequate supply of nutrients which leaves a very large yield gap of more than 40 %. Among the crops, sorghum has the largest yield gap of about 60 % with millet recording the lowest of about 30 %. Farmers in northern Ghana are of the view that sorghum does not require fertilizer for high yields and therefore do not apply fertilizer to the sorghum crop. On the other hand, the millet available are mostly local varieties and do not respond to fertilizer. With the application of fertilizers and adequate management of the millet crop, the increase in yield was just 0. 5 tons/ha compared to the rest of the crops which had increases in yield of more than 1 ton/ha.

Crop	Average yields of farmers	Achievable yields	Yield gap	Yield gap (%)
Maize	1.5	2.5	1.0	40
Rice - rainfed	1.8	3.5	1.7	49
Sorghum	1.2	3.0	1.8	60
Millet	1.0	1.5	0.5	33

Source: Ministry of Food and Agriculture (MoFA)

Table 10. National average yields (tons/ha) of some selected cereal crops in Ghana

10. Challenges to adoption of soil fertility enhacement practices

Conservation agriculture and some of the practices that enhance the fertility of soil for crop production have been tested on-station by research in Ghana and most of them have been found to be proven. These practices are now at the on-farm testing stages by research and the Ministry of Food Agriculture in different parts of the country. Some of these practices are either new to the farmers such as conservation Agriculture, Agro-forestry and Alley cropping or they are the improved versions of farmers' practices such as crop rotation (alternating cereal and legumes), intercropping cereal and legumes, root crops with cereals and legumes, cover cropping. Despite the benefits demonstrated to farmers from the use of these technologies, adoption rate is very low. Some of the challenges militating against the adoption of these practices by farmers include:

10.1. Ownership of land

Most of the farmers in Ghana do not own the land they farm on and they are therefore described as settler or migrant farmers if they come from other parts of the country and settle at that particular place. The amount of money to invest on such rented lands by these farmers will therefore depend on the length of time the land is rented for farming. A farmer with one year rent period will not be willing to invest so much on that land for farming compared to a farmer who is renting the land for over ten years. Secondly, the land owners may not even allow farmers to introduce long term investments on such short term rented lands. It will therefore be difficult for such farmers to adopt soil fertility enhancing techniques such as Agro-forestry system or even the cultivation of tree crops since this will take a long time to yield benefits to the farmer. However, It was found that the system of land tenure in the forest or the transitional zones where the farmers are allowed to use a plot for several years for farming may not have difficulties in adopting no-till as part of soil fertility management practices. This, according to Ekboir et al. (2002) if these farmers are allowed to use such lands for several years it will enable them to recoup the profits of their investments. Also, data collected on the farmers adopting any particular tillage system showed that farmers using their own lands adopted CA practices more easily than farmers on rented lands (Adjei et al. 2003).

Farmers in the forest and transition zones where share cropping arrangements exist between the settler or migrant farmer and land owners, they are encouraged to practice CA and other related practices because increase in the productivity of their crops will lead to an increase in their share of the harvest. Unlike in the north, lands are almost given out free to settler farmers and can also be seized back at anytime by land owners without any notice as there is no agreement signed between the farmers and their land owners. There are certain times land owners even seize back their lands when they find increases in crop yields of the settler farmers. In such situations, farmers in the northern Savanna zone will definitely be discouraged from adopting any of these soil fertility improving practices since the land owners do not benefit from such increases and there is also no agreement signed between them to protect the farmers from their lands being seized back.

10.2. Difficulties in maintaining soil cover

One of the three pillars CA is hinged on is the provision of adequate soil cover and it is one of the several ways of enhancing soil nitrogen. Unfortunately, it is very difficult to provide soil cover in the northern Savanna zone of Ghana because of rampant bush fires during the dry season. The northern part of the country also houses most of the country's livestock (goats, sheep and cattle). These animals either feed on the available crop residue left as a cover to the soil or the residues are removed and fed to them at home by the farmers. The removal of crop residue from the soil for livestock or the destruction by bush fires renders the soil bare. This exposes the soil to sunshine which is followed by erosion by the harmattan winds during the dry season and by running water at the beginning of the rainy season.

Most of the cover crops being promoted as materials for soil fertility improvement are not edible and so farmers are not very enthusiastic in adopting them for use. Most farmers therefore choose grain legumes among the range of soil fertility management practices due to the immediate provision of food (Chikowo et al., 2004; Adjei-Nsiah et al., 2007; Kerr et al., 2007; Ojiem et al., 2007). It has however been found that practices such as green manures and agro-forestry legumes even though do not provide immediate benefits, they are more efficient in providing nitrogen and mulch for subsequent crops (Giller and Cadisch, 1995). Experience has shown that farmers in the northern part of Ghana do not regard cover crops as part of their traditional crops and therefore cover crops have no significance in monetary terms to them. This explains why farmers are not adopting cover crops such as *Callopogonium* and *Mucuna* for soil fertility improvement even though results have shown that these cover crops give adequate cover to the soil and increase soil nutrients for high crop yields. It is difficult for farmers to replace crops they are used to with new crops especially when the new crops cannot give immediate economic returns to them.

Farmers in northern Ghana practice mixed-cropping which mostly involves cereals/legumes and cereal/cereal and this situation does not favour the inclusion of *Mucuna* and *Callopogonium* because of the climbing nature of these cover crops. Intercropping these cover crops with cereals can easily smoother the main crops and these cannot also be easily used as short season fallows since each of them needs more than six months to develop and cover the soil for optimum benefits. Work done by Loos et al. (2001) showed that *Mucuna* planted too early would result in competition for nutrients with the associated crop but planting it late also reduces its chance to properly establish. Knowing when to plant *Mucuna* is therefore very important if weed suppression and high quantity of N (150 N kg/ha) are to be achieved from the cover crop. Yield loss estimated at 30 % has also been recorded from fields of *Mucuna* intercropped with maize due to competition for nutrients, light and space.

Reports from farmers in the forest and transition zones also indicate that planting crops in no-till system is time consuming and laborious since it is by using a dibbler or a cutlass to create holes for placing the seeds. Farmers further complained that germination of seeds was negatively affected when these were planted in high amounts of soil cover. In such situations, the reduction of the soil surface mulch by partial burning becomes necessary to enhance germination which gives the farmer additional work. Other difficulties involved in planting crops under thick mulch include hidden tree stumps which could wound farmers in the process or dangerous reptiles like snakes hiding in the mulch to bite farmers during planting or weed control by hand. The introduction of jab planters reduced the time used in planting crops in the mulch but it was also dependent on the experience of the farmers.

10.3. Other uses of crop residues by farmers

Farmers in northern Ghana traditionally use crop residues as livestock feed, for housing, craft materials and as household energy source. Using crop residues as soil cover and organic matter replacement is therefore foreign and conflicts with the uses that they are already familiar with for several years. The use of crop residues such as millet or sorghum stover as livestock feed in the Upper East region of Ghana where farmers produce crops alongside rearing of animals is very common. In the Upper East region in particular, due to high population density in that part of the country, there is pressure on land use resulting into small farm lands. Despite the fact that common lands for livestock grazing is limited due to lack of land, farmers still rear livestock because of the high culture and economic value the natives place on livestock. Farm families keep livestock as investment and insurance against the risk of crop failure, for traction, for manure production and for milk and meat. These according to the farmers, make use of crop residue to feed livestock to take precedence over other uses. Animals are therefore allowed to feed on crop residues directly on the farm or the residues are carted home and fed to them. Some of the farmers are able to transport the manure from these crop residues consumed by the animals back to the fields others do not thereby depriving the soils of organic matter.

Of late, there have been increases in livestock numbers in the drier part of northern Ghana especially in the Upper East Region. The increase in the livestock industry in this region has also led to an increase for the demand for their feed and since they depend on crop residues for dry season feeding, residue for mulching the soil will be on the decrease. As a result of this intensification of livestock production, there is a fast developing market for crop residues in these areas which further encourage the removal of these residues for sale to raise income for the family at the expense of maintaining the fertility of the soil. In the northern Savanna zone of Ghana where most farmers practice mixed farming, it is therefore left for the individual either to use the crop residue for mulch or use the residues as livestock feed. So far, experience has shown that most of the farmers go in for the earlier option where they use the residues to feed their livestock. Farmers however, still have several options of improving the fertility of their soils for high yields. These include selling some of the livestock to buy in-organic fertilizers, returning the residue in the form of manure to the farm or in some cases when the plot sizes are small, compost is produced and applied to the crops.

Experience has also shown that in areas where livestock numbers are low such as in the forest and transition zones, the crop residues are not left as mulch but burnt as practiced in the traditional slash and burn system of farming. Apart from burning to control weeds in that system, it has also been found that burning helps to control pests and to reduce the population of rodents which tend to increase when crop residues are left on the field. Even though retention of crop residues is always advocated in CA, under situation of very high mulch content, retention is not feasible and burning to reduce it seems to be a good option.

10.4. Weed management

At first sight, spraying to kill the existing vegetation in the northern Savanna zone to plant a crop as in CA system appears like no other weed will ever germinate again. However, two weeks after planting the crop, one finds a huge mass and diversity of weeds vigorously springing up thereby making the first weeding after planting very difficult and laborious since this is done by hand. At times the high infestation of weeds in such a system is due to bad selection of weedicide, low doses of the chemical and poor spraying techniques.

Rio (1992) estimated over 45 % as the annual yield loss of crops due to weed infestation in heavily infested fields. Other effects include waste of human energy in controlling weeds. It has been found that reducing tillage intensity alone as described in CA without adequately covering the soil as practiced by most farmers is one way of promoting weed infestation on their fields. In the situation whereby crops are planted haphazardly leaving some gaps, weeds quickly infest and occupy these areas making their control very difficult as this requires weeding several times by hand hoe.

10.5. Unavailability of cover crop seeds

Even though some farmers now know the benefits of growing cover crops as an intercrop with their cereals or in a short fallow system, seeds of these cover crops are not easily available. Few of the cover crops which have been tested and found useful for soil fertility enhancing materials are *Mucuna, Pueraria, Canavalia* and *Callopogonium species* but their seeds are scarce and difficult to find in Ghana. This makes it difficult for many farmers to adopt and use these materials to enhance the fertility of their soils. It has been observed that seeds of these cover crops are easily found when there is a project promoting them. For example when there was intensive promotion of *Mucuna* for weed suppression and soil fertility in the transition zone of Ghana, it was easy to get *Mucuna* seeds to buy because some farmers produced the seeds for sale. Initially, the project motivated the farmers to produce the seeds by buying the seeds off from them during the project but when the project ended and seeds were not bought again, the farmers also stopped producing the seeds. This situation calls for the introduction of edible cover crops because if *Mucuna* were edible, farmers would have continued to produce the crop for food while maintaining the fertility of their soils.

Cover crops which are not edible as mentioned earlier are less attractive to farmers because they do not give immediate benefit to the farmer. The option of growing cover crops as short season fallows is more feasible in the transition zones where population is less dense with large farms but not in areas where there is pressure on land use and there is no fallow period permitted. Practicing no-till on bare soil with less than 10 % surface mulch as in the Savanna zone with rampant bush burning may result in reduced crop yields.

10.6. Difficulties in getting appropriate equipment and tools

The practice of some of the soil fertility enhancing technologies such as in Agro-forestry and CA require the use of some equipments, inputs and tools. It has been observed that the required inputs such as glyphosate as used in CA land preparation are mostly not available at the appropriate time needed by farmers. At times where they are available the cost may be so high that the average farmer may not be able to afford. Agro-inputs distribution is therefore described as being poor because the right inputs are always not available at the right time needed. Chemical fertilizers which are needed to generate biomass at the beginning of CA practice in the Savanna zone have their depots located in the regional capitals of the country making it almost impossible for most of the farmers who are in rural areas. One of the difficulties involved in the implementation CA and other technologies designed to enhance crop yields may be lack of accessibility to inputs such as weedicides.

The Food Crops Development Project (financed by the African Development Bank) implemented mostly in the forest and transition zones saw the supply of inputs to project clients by some major agro-input companies. According to Boahen et al., (2007) farmers use the project as collateral to gain access to the inputs for crop production. In that system the Project sent a request to the shop to provide a certain quantity of inputs, for which the farmer pays later into a bank account created for this purpose. The Inputs and tools supplied to the farmers in this system ranged from pesticides and fertilizer to equipment and tools like knapsack sprayers, cutlasses and hoes. The above indicates that the implementation of any soil fertility enhancing technologies in Ghana by donors through projects is always successful but the systems breaks down immediately the project ends.

This calls for sustainability to be built into every project implemented to make sure that the farmers own and operate the system even after the project. Some of the suggestions given to introduce sustainability into such projects include:

Urging the farmers to form co-operatives where they can be registered and linked to financial institutions such that even after the project, the farmers can get financial assistance from such Institutions

Secondly, training of farmers on both the process and the content of the project will be very important for the visibility of the project after it has been concluded.

The donors or project implementers should always look for their local partners and work with them. This will enable the activities of the project to continue through the efforts of the local partners after the duration of the project.

Farmers in the north of Ghana have expressed their gratitude for the introduction of CA and other related technologies as labour is scarce and some of these practices are ways of reducing labour costs for crop production. Technologies such as no-till and direct seeding are practices demonstrated to the farmers of the north which are devoid of tilling the land either by hand hoe or tractors. Even though the introduction of these technologies are appropriate, the equipments and tools to go with these practices are not yet available for sale in Ghana except those used for the demonstrations. The planting of the crops in the mulch without these tools remains a challenge to the adoption of these technologies as farmers spend more time to get the seeds planted using dibblers and cutlasses. Implements like knife-rollers, rippers and no-till seeders are needed to facilitate planting in CA which are yet to be made available in the country. According to the farmers, planting is easier with dibblers or cutlasses when the soil is bare but becomes more difficult if there is a high surface mulch as in CA and the crop is to be planted in rows.

Currently in Ghana, the most common practice available for medium- and large-scale farmers is the tractor-mounted disc plough and harrow which are imported and sold to either individuals or group of farmers. However, due to the fragile nature of the Savanna soils, these equipments have been observed to be responsible for the destruction of the soil structure and increase soil erosion by running water during the rainy season. So far, the components of CA being demonstrated are targeting the small-scale farmers but if it is to be adopted by the medium and large-scale farmers, the availability of machines and equipment becomes very necessary. In order to expose the technologies to these category of farmers, there is the need to develop appropriate machinery, tools and other implements or at best adapt the existing ones, fabricate them and make them accessible to such farmers. This can be done by effectively training the local artisans and craft men/women to produce such equipments for the farmers. Even though most of these equipments are operated by tractors, if those to be produced are designed to be operated by bullocks and donkeys, it would attract many more farmers to adopt the practice.

10.7. Farmers lacks access to credit and markets

Farmers in northern Ghana have been known to be poor probably because of the poor harvest they obtain from their crops which is traced mostly to low soil fertility and erratic rainfall. Most of them therefore lack collateral security to obtain financial assistance from these financial Institutions. Meanwhile the adoption of any soil restoration practices such as CA requires the purchase of inputs such as weedicides and fertilizers and other equipments for direct planting and spraying of the weedicides. The inability to purchase these inputs therefore means that such farmers would not be able to adopt such soil fertility enhancing practices.

The system where farmers sell their crop produce through the middlemen is one of the reasons why most of them remain poor. Prices offered to the farmers by these middlemen for their produce are so low that they are never able to pay for the cost of production. Farmers whose activities for crop production are pre-financed by these middle men suffer most as they take the produce in lieu of cash at harvesting time when prices are generally low. In addition such farmers might not easily adopt some soil fertility techniques such as CA which does not give immediate returns to the farmer.

10.8. Pests and disease problems

Experience has shown that providing thick soil mulch creates a micro-climate for reptiles such as snakes that can bite farmers operating on the land without protective clothing. Most farmers also complained of the increase in scorpions and other insects which can cause significant losses to the crops planted in the mulch. Some farmers who were introduced to direct seeding in the mulch complained of poor plant stand due to damage by crickets and grass hoppers. According to the farmers, that is why they have cultivated the habit of burning the crop residues before planting. It is now well known by farmers who have ever produced cover crops such as *Mucuna* that the cover crop is a suitable abode for snakes both in the live stage and when it is dry and left on the soil as mulch which makes farm operations by farmers very dangerous. Some farmers have come out with a calendar of spraying programme to control the pests in these cover crops to save the crops from insect damage and to make farm operations safe. This is because the chemicals (insecticides) used in spraying to control the insects also drive the snakes and the scorpions out of the mulch. However, the farmers are of the view that the adoption of such technologies also increases cost of production since they have to purchase insecticides to control these insects.

10.9. Difficulties in promoting soil fertility enhancing technologies

In the first place, in Ghana the number of extension agents of the Ministry of Food and Agriculture (MoFA) responsible for agricultural extension services is very low. This makes it impossible for them to have a large coverage of farmers within a specific time to effectively extend whatever new technology developed by Research. Also, the knowledge of these extension agents in the various soil fertility restorations may be low compared to other subject areas. It has been realized from experience that the knowledge and lessons learnt from the past soil fertility enhancement project have not been made use of by the agents indicating that they have not been trained in that line.

Until recently, MoFA was structured into departments such as crop and extension services departments and for soil fertility restoration technologies to be extended effectively for adoption, there is need to establish a unit to champion the activities of this subject which is neither crops nor extension services.

So far, areas with conservation Agriculture and its related practices have been traced to the existence of donor projects. The donors of all these projects have been in collaboration with MoFA with the activities carried out since the extension agents have been responsible for site and farmer selection. However, Boahen et al. (2007) reported that the number of farmers using these technologies reduced by an estimated 30% when the related projects ended. It was also found that the visits by extension agents to these project communities reduced from twice a week to once every two weeks since the project was no longer giving them fuel allowances and the associated cost of travel.

10.10. Lack of adequate institutional support

For farmers to adopt the soil fertility regeneration and maintenance practices, its extension needs to be well co-ordinated and collaborated among major stakeholders in soil health. In that way, the numerous benefits of CA can be realized and appreciated by farmers, researchers and extension staff of MoFA. So far, even though the activities of these projects have been carried out in collaboration with MoFA, data on the practice and adoption of these including CA are scanty. It has been observed that even where data exists, they may not be coherent or accurate. There are few success stories on some of the practices that can regenerate and maintain soil fertility for crop production but these are not properly packaged for extension and for policy makers. Both donors and the implementers of CA in Ghana have therefore not been able to convince policy makers the benefits of these practices for support. This may explain why the Government still imports tractors with both disc ploughs and harrows without considering the importation of equipment and tools used in CA. The adoption of CA and its related practices by farmers does not really depend on the availability of these tools and inputs; it also depends on the attitudes of all stakeholders in the supply chain such as the input dealers and manufacturers.

The Government of Ghana's modernization of agriculture programme through the Ministry of Food and Agriculture seeks to modernize and mechanize agriculture in Ghana. This programme has begun with the importation of modern tractors equipped with new disc ploughs and harrows probably targeting the medium and large scale farmers with the neglect of the small scale farmers of less than one hectare. This situation also makes it difficult for even the donors to fund such soil fertility regeneration practices like the CA because it would be like working directly against the government's programme. A compromise could have reached by the importation of tractors and no-till seeders and other equipments for the small scale farmer who can not afford the services of the tractor ploughing services.

10.11. Use of farmers' indigenous knowledge with the technology

The activities of research and development in producing any agricultural technology for the farmer should be seen to be improving or incorporating farmers' indigenous knowledge and not producing modern technology that seem not to have any input from them. The adoption of any technology developed in a participatory manner among the farmer the researcher and the extension agent seem to be faster than when indigenous knowledge from the farmers are ignored. This situation calls for in-depth studies on traditional practices and the strategies farmers employ to cope with the declining soil fertility status of the savanna region. This will assist in the development of a technology that will not be difficult to extend to the farmers for adoption if the technology is built on the existing indigenous knowledge of the farmers. The introduction of the use of herbicides to kill the weeds for the planting the crop was met with happiness in some parts of the northern region of Ghana where farmers plant on the bare soil immediately after the onset of rains without tilling the soil. In another development, earth and stone bunding to control both soil and water has not been difficult to extend the technology for adoption by farmers with their farms located on steep slopes and rocky areas because they were already doing something similar before. Most of the farmers are not adopting the cultivation of cover crops such as *Mucuna* because the crop is not edible. It is envisaged that if the edible cover crops such as the local creeping cowpea is introduced to replace *Mucuna*, a lot more farmers will adopt the practice as both the leaves and grain of the cowpea are edible.

11. Conclusion and recommendations

From experience, conservation agriculture and other related practices geared towards regenerating soil fertility for crop production had many important impacts on the lives of adopters. Practices such as Agro-forestry, use of cover crops and many more have helped in increasing grain yields of crops several folds. Also reducing the number of times the soil is tilled as in CA has helped to reduce energy and labour costs and the decay of the vegetative mater has improved soil fertility status which in several places has translated into high crop yields.

From the numerous works done in the areas of soil fertility management and other related field activities carried out in the past, showed that conservation technologies in general, is site-specific and depends on the local bio-physical and socio-economic settings. Also from the interaction with farmers, there are many important changes that CA brought to farming activities, they mentioned reduced investment in cash and labour, higher yields, easier weed and pest control, and saved time for farmers.

From the knowledge gained in the past from several desk tops research and other methods of research conducted to assess the extent which CA and other soil fertility enhancing practices have been adopted by farmers in Ghana, the following observations and conclusions have been drawn:

Farmers adopting CA in the savanna region of Ghana are faced with challenges of generating enough biomass to begin with and the control of weeds in the transitional phase of the system. Related to weed control in CA is the knowledge of the type of herbicide to use and mode of application for effective control of the emerged weeds.

In order to effectively extend CA and other practices related to soil fertility regeneration for adoption, enough information especially on cover crops and their profitability are needed. Also knowledge on how they fit into the various ecologies of Ghana, the best crop associations possible and their effects on soil fertility status, are essential for the dissemination of the cover crop technology among farmers.

It is essential to have relevant knowledge in the selection of herbicides and mode of application. The selection and timely incorporation of cover crops in the various cropping systems are also important in the practice of CA. Some of the skills required by farmers practicing CA are the ways and means of controlling rodents and other pests in order to obtain high yields.

In the Savanna zone of Ghana, it is difficult to maintain soil cover with crop residue or cover crops because of the rampant bush fires during the long dry season and the grazing of several different livestock on free range.

Tools and inputs such as seeds of appropriate types of cover crops, the required herbicides to be used in CA practices are essential requirements for the practice of CA which may not be easily available to farmers at the right time.

Some of the requirement for the practice of soil fertility regeneration technologies such as agro-forestry and growing of cover crops in improved fallows system help to raise the total cost of production. This is because these require initial land clearing in the case of the cover crop fallow, additional labour for spreading the mulch and planting through the residue, which is laborious and time wasting.

Even though direct planting without tillage in CA is said to be cheaper than conventional land preparation, due to the scarcity of herbicides at planting time, the price of the commodity is high enough to cause significant impact on cash demand of farmers during the farming season.

Manual planting by hand, using stick or cutlass on fields with mulch is more difficult and time wasting than using the same tools to plant on bare or conventionally ploughed fields.

It is important to impact basic knowledge in handling of equipment like spray machines for herbicides or application of other chemicals to farmers. Farmers' yields can be enhanced if they are assisted in decision making on appropriate crops to be grown, rotations, record keeping and costing of each operations.

Following the observations and findings from our experiences on soil fertility management for crop production, the following recommendations are made for further observation and consideration:

Depending on the availability of ready market, emphasis on cereal-legume rotations/intercropping for CA should involve multipurpose grain and fodder legumes. The production of fodder in the system will take care of the livestock component since on the average every farm family rears animals alongside crop production in the Savanna zone of Ghana.

Training farmers on aspects of CA and introducing to them simple and appropriate CA equipment and implements will significantly enhance labour productivity and encourage many more farmers to adopt CA and other related soil fertility improvement practices.

In order for the Donor-led projects to build a good number of success stories on the various soil fertility management practices including CA, the knowledge and experience acquired over the past years should be harnessed, repackaged and used. In this way, they will be able to convince the government on the benefits, of CA and its potential to resolve food security problems and promote a sustainable source of livelihood for rural small-scale farmers.

The promotion of most of these practices of CA should go beyond the small scale farmer since there are emerging medium scale farmers who can help broaden the scope of the programme. This can be done through adaptive research targeting different groups of farmers in the different environments and socio-economic settings.

The first step in promoting CA effectively is to consider National institutions and farmers' groups to be the driving forces of CA in the country. These groups can re-package and properly lobby the policy regularly for support.

Messages developed for extension services should be specific for the occasion and not blanket for all issues as it is now. This will assist the farmers to be able to assess their constraints and be able to opt for suitable practices that can maintain and improve soil fertility for sustainable crop production.

Both farmers and extension officers need to understand the processes through which the soil fertility can be maintained for extension to enable farmers to adopt if the practices are imported from other regions and not for adaptation alone. This is important because if the environmental factors of the place the practice is imported from are the same with the local environment, but the socio-economic setting may be different.

Intensification of the integration of cover crops and crop rotation in CA systems should be pursued and monitored since there are several cover crops and other crop varieties. This situation can lead to the reduction of pests and diseases in the CA systems.

Farmers should be encouraged to form co-operatives. This is because if the groups are well co-ordinated, agricultural policies including the regeneration of soil fertility for crop production can be well implemented

Author details

James M. Kombiok[1], Samuel Saaka J. Buah[2] and Jean M. Sogbedji[3]

*Address all correspondence to: kombiokjm@yahoo.com

1 Savanna Agricultural Research Institute, Tamale, Ghana

2 Savanna Agricultural Research Institute, Wa Station, Wa, Ghana

3 IFDC, Rue Soloyo, Lome, Togo

References

[1] Adjei E.O., Aikins, S.H.M., Boahen, P, Chand, K, Dev, I., Lu, M., Mkrtumyan, V., Samaraweera, S.D., and Teklu, A. 2003. Combining mechanisation with conservation agriculture in the transitional zone of Brong Ahafo, Ghana. International Center for Development-Oriented Research in Agriculture, Working Document Series 108. Wageningen, Netherlands: ICRA.

[2] Adjei-Nsiah, S., Kuyper, T.W., Leeuwis, C., Abekoe, M.K., Giller, K.E., 2007. Evaluating sustainable and profitable cropping sequences with cassava and four legume crops: effects on soil fertility and maize yields in the forest/savannah transitional agro-ecological zone of Ghana. Field Crop Res. 103, 87–97.

[3] Adu-Gyamfi, J.J., Myaka, F.A., Sakala, W.D., Odgaard, R., Vesterager, J.M., Hogh-Jensen, H., 2007. Biological nitrogen fixation and nitrogen and phosphorus budgets in farmer-managed intercrops of maize–pigeon pea in semi-arid southern and eastern Africa. Plant Soil 295, 127–136.

[4] Agyare, W., J. M. Kombiok, N.N. Karbo, & A. Larbi (2002). Management of pigeon pea in short fallows for crop-livestock production systems in the savanna zone of northern Ghana. Agroforestry Systems, 54: 197-202

[5] Boahen, P., Dartey, B.A., Dogbe, G.D., Boadi, E.A., Triomphe, B., Daamgard-Larsen, S. Ashburner, J. 2007. Conservation agriculture as practiced in Ghana. African Con-

servation Tillage Network, Centre de Coopération Internationale de Recherche Agronomique pour le Développement, Food and Agriculture Organization of the United Nations, ISBN :9966-7219-1-6.

[6] Buah S. S.J. 2004. SARI Annual Report 2004. Savanna Agricultural Research Institute annual report – Upper West farming systems research group.

[7] Carsky R.J., Becker M. and Hauser S. 2001 Mucuna cover crop fallow systems: potential and limitations. In: sustaining soil fertility in West Africa, Tian G., Ishida F. and Keatinge J.D.H. (ed.). SSSA Special publication no 58, Soil Science Society of America and American Society of Agronomy, Madison, WI, USA. pp. 111-135

[8] Chikowo, R., Mapfumo, P., Nyamugafata, P., Giller, K.E., 2004. Woody legume fallow productivity, biological N2-fixation and residual benefits to two successive maize crops in Zimbabwe. Plant Soil 262, 303–315.

[9] Ekboir, J., Boa, K. and Dankyi A.A. 2002. Impacts of No-Till Technologies in Ghana. Mexico, D.F.: CIMMYT. pp32.

[10] Ekekpi, G.K and Kombiok J.M. 2008. Report on baseline study in three target districts of conservation agriculture project in Northern Ghana. Care International Report.

[11] [EPA] Environmental Protection Agency. 2003. National action programme to combat drought and desertification. EPA.

[12] FAO, 2008. Conservation Agriculture. 2008-07-08 http://www.fao.org/ag/ca/index.html.

[13] Fosu M., Kühne R.F. and Vlek P.L.G. 2004. Improving maize yield in the Guinea savanna zone of Ghana with leguminous cover crops and PK fertilization. J. of Agron. 3 (2): 115-121.

[14] Giller, K.E., Cadisch, G., 1995. Future benefits from biological nitrogen-fixation—an ecological approach to agriculture. Plant Soil 174, 255–277.

[15] Horst W.J. and Härdter R. 1994. Rotation of maize with cowpea improves yield and nutrient use of maize compared to maize monocropping in an alfisol in the northern Ghana savanna of Ghana. Plant and Soil 160:171-183.

[16] Kerr, R.B., Snapp, S., Chirwa, M., Shumba, L., Msachi, R., 2007. Participatory research on legume diversification with Malawian smallholder farmers for improved human nutrition and soil fertility. Exp. Agric. 43, 437–453.

[17] Kombiok, J. M. and Clottey, V. A. 2003. Maize yield and soil N as affected by date of planting Mucuna in maize/mucuna intercropping system in Ghana. Trop Agric. 80 (2): 77-82.

[18] Kombiok, J. M., Rudat H., Frey E. 1995. Effects of short term Callopogonium fallow and different soil management practices on yield of maize in Northern Ghana. In: V. Akita, P. Schroder and S. K. Bemile (Eds.). Organic and Sedentary Agriculture pp 1-5.

[19] Loos H. 2001. Agricultural mechanization in Ghana, Technical and organizational options. Paper presented in workshop, 'Mechanization of agriculture: missing link to agro processing', Crops Research Institute, 19–20 December 2001, Kumasi, Ghana.

[20] Loos H, Zschekel W, Schiller S, Anthofer J. 2001. Integration of Mucuna improved fallow systems into cropping systems of the Brong Ahafo Region. Paper presented at international conference organized by the Soil Science Society of Ghana, 26 February–3 March, Tamale.

[21] MOFA (2002). Food and Agriculture Sector Development Policy (FASDEP), Government of Ghana. Delaram Ltd. Accra. 68pp.

[22] Ojiem, J.O., Vanlauwe, B., de Ridder, N., Giller, K.E., 2007. Niche-based assessment of contributions of legumes to the nitrogen economy of Western Kenya smallholder farms. Plant Soil 292, 119–135.

[23] Quansah, C., Osei-Yeboah, S., and Osei-Bonsu, P. 1998. Assessment of the implementation of village land development plans – Adoptive trials and demonstrations. Land and Water Management Unit. Ministry of Food and Agriculture, Accra. Ghana'

[24] RELC, NR. (2004). Outcomes of Northern Region RELC Planning Session for 2004. Tamale.

[25] RELC, NR. (2005). Outcomes of Northern Region RELC Planning Session for 2005. Tamale.

[26] SARI (1995). Savanna Agricultural Research Institute (SARI) annual Report for 1995, Nyakpala, Ghana.133pp.

[27] SARI (2005). Savanna Agricultural Research Institute (SARI) annual Report for 2005, Nyakpala, Ghana.105pp.

[28] Schmidt G. and Frey E. 1992. Cropping systems research at the Nyankpala Agricultural Experiment Station. In Improving farming systems in the interior savanna zone of Ghana. Acquaye and NAES (ed.) pp 14-35.

[29] Wagger, M.G. and Denton, H.P. 1992. Crop and tillage rotations: Grain yield, residue cover and soil water. Soil Sci Soc. Am. J. 56:1233-1237

Improving Fertilizer Recommendation and Efficiency

Chapter 4

Plant Analysis

Renato de Mello Prado and Gustavo Caione

Additional information is available at the end of the chapter

1. Introduction

Several tools are available to evaluate the nutritional state of plants. Plant analysis is an efficient one since it uses the plant itself as a nutrient extractor. Thus, it complements soil chemical analysis and makes it possible to predict nutritional disorders before the appearance of visual symptoms in the plant tissue. However, it is necessary to integrate both techniques, chemical analysis of plants and chemical and physical analysis of soil, besides visual diagnosis to maximize fertilization efficiency in terms of cost and prevention of environmental damage.

Adequate fertilization avoids damage to the environment by reducing soil acids, water euthrophy, pollution of the phreatic zone and area salinization. Furthermore, efficient fertilizer handling is fundamental in any productive system, especially in the recent decades, due to increased cost, scarcity of some nutrient sources and consumer insistence for high quality products.

Precise analytic methods only are not sufficient to an adequate fertilization handling. A competent professional having theoretical and practical experience and knowledge about the various factors involved in the production chain, like interactions "soil-plant-environment- handling, is also an absolute requirement.

By plant analysis it is possible, among others, to determine culture nutrient needs and exportation, identify nutritional deficiencies that produce similar symptoms, evaluate nutritional states, help in the managing of fertilization programs and diagnose about levels of nutrients in diverse plant organs. Several procedure, direct and indirect, are available to achieve these aims. This chapter will emphasize the main methods utilized in the diagnosis of the nutritional state of plants, like chemical foliar analysis, biochemical tests, measurements of leaf green color and visual observation. The linked information in the present chapter were obtained from an extensive literature review and also were inserted professional experiences of the authors.

Again, it is emphasized that efficient foliar diagnosis includes all procedures starting from correct field sampling to adequate laboratory analysis.

2. Factors that affect nutritional diagnosis

It is important to know the main factors that interfere in the diagnosis of plant nutritional status, so that confident analytic results are obtained and compared to pre-established literature standards. The composition of the vegetal tissue reflects the interaction of factors acting up to the moment the samples are collected for analysis.

Initially, it is necessary to exclude biotic and abiotic factors that affect nutrient concentrations in plants. Among these should be considered, lack or excess of water, high or low temperatures winds, pests and illnesses, compacted or poorly plowed soils, mechanical damage and herbicide toxicity [1]. These and other factors may produce deficiency symptoms in the plant by preventing absorption and/or translocation of nutrients. In such cases symptoms of deficiency are only eliminated by removing the stress factors.

Table 1 shows some of the factors directly involved in the appearance of real or apparent symptoms, which are similar and confound typical deficiency and toxic patterns.

Factors	Cause
Biotic	Occurrence of illnesses and/or pests causing damage to the aerial or root systems in the plant and inducing symptoms similar to mineral deficiencies.
	Natural leaf senescence produces color changes.
	It is a tightly regulated process involving the coordinated expression of specific genes and hormonal participation, mainly cytokinins and ethylene in sequential events and mechanisms that are not well known [2].
Abiotic	Extreme environmental conditions (temperature, drought, floods, strong winds), especially in the 10 to 15 days before sampling and recent removal of weeds.
	Inadequate application of products or interaction of products like chemical fertilizers, organic matter, fungicides, insecticides herbicides, antibiotics, growth regulators or foliar fertilizers, which could prevent absorption of a nutrient and/or simulate deficiency symptoms.
	Inadequate physical and/or chemical soil conditions like poor manipulation, erosion, sharp slopes, excess aluminum, iron or manganese, low levels of available nutrients.
	Culture practices inductive of plant abnormal symptoms, like poor irrigation, addition of organic matter not completely digested, intense pruning and deep soil harrowing.

Table 1. Effects of biotic and abiotic factors, that may directly or indirectly induce typical deficiency or toxicity patterns. (Adapted from [1]).

To exclude the factors in Table 1, the professional must know the interactions "plant-environment-soil – farming activity" before proceeding to sampling.

3. Criteria for plant sampling

In this item, sampling criteria will be discussed, in terms of leaf analysis, although it should be emphasized that these principles are also applied in visual diagnosis and indirect methods.

Sampling is a fundamental step in the outcome of foliar analysis. Poor or inadequate sampling compromises all available recommendations. Results quality and precision are directly dependent on the procedure. This is a critical step since nutrients concentrations are not the same in all plant parts, and may differ according to age and variety. Foliar analysis results will only be useful and representative of the culture if sampling is correctly performed.

Some criteria are similar to the ones employed in soil sampling and follow basic procedures [3].

1. Cultures should be divided in plots not bigger than 10 ha, having uniformity in age, variety, spacing, soil and manipulations.

2. In each plot, the indicated leaves from the desired cultures are collected in a zigzag direction.

3. Preferably, collections should be made between 7 and 11 AM, more than 24 hours after a rain

4. At least 20 leaves must be collected from each plot and mixed before being sent to the laboratory.

5. All samples must be packed in clean unused paper bags to avoid contamination.

6. Samples are identified by tags corresponding to each plot

7. Samples should be immediately sent to the laboratory. When this is not possible the material must be kept in an isolated container, fitted with a 150w lamp during 72hours, for initial drying.

8. Sampling must never be conducted after fertilization or spraying. In these cases, collection of samples is made 30 days later to avoid foreign residues.

9. Leaf samples are sent to the Foliar Analysis Laboratory after complying with the rules described.

Additional important details are:

- damaged or abnormal looking leaves must not be collected unless this is caused by nutritional problems

- soil-contaminated samples should be avoided and also the ones collected from plants situated close to roads or entrance pathways.

- sample collector must make sure hands are clean.

- samples packed in open and perforated paper bags sent to the laboratory two days after collection do not need decontamination and previous field drying procedures. If this is not possible, samples could be treated as already described or a) washed successively in clean water, 0.1% detergent solution, clean water followed by drying in a 70ºC oven or in a sunny environment before being sent to the laboratory and b) samples packed in polyethylene bags may be kept at low temperatures (2 to -4ºC) for a maximum of 72 hours.

Recent matured leaves are the usual plant organ analyzed but eventually stem pieces or branches may be used. In leaves, analysis may be performed in the whole structure or only in specific parts like the lamina or the petiole. In some cases, like in sugar cane, the leaf midrib is removed when foliar diagnosis is desired. In perennial cultures, like coffee or citrus, leaf composition may vary by the presence or absence of fruit in branches. In general, recent matured and physiologically active leaves are the plant organs, which better reflect the nutritional status. They respond more readily to variations in nutrient supply and are, thus, better qualified as samples.

Concerning the number of samples, it must be enough to reduce variability and be representative of the plant population. In rare occasions, dry material in each sample must exceed 10g (100 to 200g fresh green tissue for most species), but this indicates that different number of samples may be necessary for particular needs of cultures and soils. On average, it is considered that 20 single units would be sufficient to compose a sample [1].

4. Recommendations about foliar sampling and adequate nutritional levels in some plant species

To evaluate nutritional status the sample, one plant, a set of plants or a previously determined plant part, must be compared to a standard, which consists of a set of nutritionally "healthy" plants. A plant is considered "healthy" when all its tissues show nutrients in adequate quantities and proportions, it is able to attain high productivity and it looks like the specimens found in very productive cultures. However, the reference culture must be as close as possible of the culture to be sampled and analyzed, and it should be a true representative of the peculiar soil-climate characteristics as well as of type of handling and ecological zoning. The reference may have the best productivity but the comparison must be with the same genetic material under the same handling regime and the sampling must follow the same procedures for normal and problem plants.

It is important to establish which plant part it is going to be analyzed in the best period since composition of different parts is not the same and nutrient concentrations also vary according to growth stage.

The previously established physiological stage of the comparison standard must be kept if it is available, or else if this is not existent the start of the reproductive stage should be preferred being a period of the highest nutrient concentration. Thus, if a deficiency is detected, it can still be corrected and it will not compromise or minimize the productivity of the next crop.

The sampling must follow recommendations, as discussed above, to produce reliable analytical results that will be compared to a standard. The analytical results should be produced by a competent laboratory engaged in a constant quality control program.

Publications by several authors, [4], [5], [6], [7], [8], [9], [10], [11], report previously defined plant organs, sample numbers and the sampling period for diverse cultures. But to utilize such data as a standard it is necessary to be careful about the physiological age of each plant and leaf as stated by the author.

Table 2 shows that there is not a standard method of sampling for all cultures. Furthermore, adequate levels of nutrients vary according to different authors emphasizing the care that should be taken to always consider the same author when following a method of collection and in comparing the adequate nutrient levels. It should be noted that adequate levels of chlorine were not described and only some gave values for molibdenium.

Culture	Plant organ; number of samples and period of sampling	Adequate dosages		References
		Macronutrients (g kg⁻¹)	Micronutrients (mg kg⁻¹)	
	--- Fruits ---			
Avocado	50 leaves (1/plant) for each homogeneous plot. Type of leaf: leaves 5 to7 months old, recently expanded from medium height crowns. February to March.	N - 16-20 P - 0,8-2,5 K - 7-20 Ca - 10-30 Mg - 2,5-8,0 S - 2,0-6,0	B - 50-100 Cu - 5-15 Fe - 50-200 Mn - 30-100 Mo - 0,05-1,0 Zn - 30-100	[4]
Pineapple	50 leaves (1/plant) for each homogeneous plot. Type of leaf: recently matured "D" (generally the 4th leaf from the apex), soon before floral induction. Cut leaves in pieces of 1 cm wide, eliminating the basal portion without chlorophyll. Homogenize and separate about 200 g to be sent to the laboratory.	N - 15-17 P - 0,8-1,2 K - 22-30 Ca - 8-12 Mg - 3-4 S - 2-6	B - 20-40 Cu - 5-10 Fe - 100-200 Mn - 50-200 Mo - Zn - 5-15	[6]
Acerola Barbados Cherry	50 leaves (1/plant) for each homogeneous plot. Type of leaf: to sample the 4 sides of the plant, for young leaves totally expanded from fructifying branches.	N - 20-24 P - 0,8-1,2 K - 15-20 Ca - 15-25 Mg - 1,5-2,5 S - 4-6	B - 25-100 Cu - 5-15 Fe - 50-100 Mn - 15-50 Mo - Zn - 30-50	[4]
Banana	25 leaves (1/plant) for each homogeneous plot less than 4ha. Tree it is recommended to sample the third	N - 27-36 P - 1,6-2,7 K - 32-54	B - 10-25 Cu - 6-30 Fe - 80-360	[4, 5]

Culture	Plant organ; number of samples and period of sampling	Adequate dosages		References
		Macronutrients (g kg⁻¹)	Micronutrients (mg kg⁻¹)	
	leaf from the apex when the inflorescence shows all the uncovered female bunches (without bracts) and not more than three male flower bunches. Collect 10 to 25 cm of the internal median part of the limb and eliminate the central rib. For varieties: Nanica, Nanicão e Grande Naine,under irrigation regimens.	Ca - 6,6-12 Mg - 2,7-6 S - 1,6-3	Mn- 200-1800 Mo - Zn - 20-50	
Banana	30 leaves (1/plant) for each homogeneous plot. Type of leaf: the 5-10 cm central part of the 3ʳᵈ leaf from the inflorescence, eliminating the central rib and the peripheral halves.	N - 27-36 P - 1,8-2,7 K - 35-54 Ca - 3-12 Mg - 3-6 S – 2,5-8	B - 10-25 Cu - 6-30 Fe - 80-360 Mn- 200-2000 Mo - Zn - 20-50	[6]
Banana	25 leaves (1/plant) for each homogeneous plot less than 4ha. For the banana tree it is recommended to sample the third leaf from the apex, when the inflorescence shows all the uncovered female bunches (without bracts) and not more of three male flower bunches. Collect 10 a 25 cm of the internal median part of the limb, and eliminate the central rib. For varieties: Prata, Anã, under irrigation regimens	N - 25-29 P - 1,5-1,9 K - 27-35 Ca - 4,5-7,5 Mg - 2,4-4,0 S - 1,7-2	B - 25-32 Cu - 2,6-8,8 Fe - 72-157 Mn - 173-630 Mo - Zn - 14-25	[11]
Orange	100 leaves (4leaves/tree), for each homogeneous plot. Type of leaf: 3ʳᵈ leaf from the fruit. Leaf born in the spring, 6 months old, in branches with fruit 2 to 4cm in diameters.	N - 23-27 P - 1,2-1,6 K - 10-15 Ca - 35-45 Mg - 2,5-4,0 S - 2-3	B - 36-100 Cu - 4-10 Fe - 50-120 Mn - 35-300 Mo - 0,1-1,0 Zn - 25 - 100	[6]
Fig	100 leaves (4 leaves/tree) for each homogeneous plot. Type of leaf: recently matured and totally expanded leaf, in the middle portion of a branch, 3 months after sprouting.	N - 10-25 P - 1,0-3,0 K - 10-30 Ca - 30-50 Mg - 7,5-10 S - 1,5-3,0	B - 30-75 Cu - 2-10 Fe - 100-300 Mn - 100-350 Mo - Zn - 50-90	[4]

Culture	Plant organ; number of samples and period of sampling	Adequate dosages		References
		Macronutrients (g kg⁻¹)	Micronutrients (mg kg⁻¹)	
Guava c.v. Paluma	30 leaves (1/plant) for each homogeneous plot. Type of leaf: 3rd pair of leaves, recently matured (with petiole) from branch extremities, collected in the period of full bloom in the culture.	N - 20-23 P - 1,4-1,8 K - 14-17 Ca - 7-11 Mg - 3,4-4 S - 2,5-3,5	B - 20-25 Cu - 20-40 Fe - 60-90 Mn - 40-80 Mo - Zn - 25-35	[7]
Mango	80 leaves (4/tree) for each homogeneous plot. Type of leaf: middle leaves in branches with flowers in the extremities from the last vegetative flux. Thus, during florescence.	N - 12-14 P - 0,8-1,6 K - 5-10 Ca - 20-35 Mg - 2,5-5 S - 0,8-1,8	B - 50-100 Cu - 10-50 Fe - 50-200 Mn - 50-100 Mo - Zn - 20-40	[6]
Apple	100 leaves (4 a 8/plant) for each homogeneous plot. Type of leaf: recently matured and totally expanded.	N - 19-26 P - 1,4-4 K - 15-20 Ca - 12-16 Mg - 2,5-4 S - 2-4	B - 25-50 Cu - 6-50 Fe - 50-300 Mn - 25-200 Mo - 0,1-1,0 Zn - 20-100	[6]
Papaya	15 petioles of young leaves, totally expanded. (1/tree) for each homogeneous plot. When leaves are mature (17th to 20th leaves from the apex), with a visible axially set flower.	N - 10-25 P - 2,2-4 K - 33-55 Ca - 10-30 Mg - 4-12 S -	B - 20-30 Cu - 4-10 Fe - 25-100 Mn - 20-150 Mo - Zn - 15-40	[6]
Passion fruit	20 laves (1/tree) for each homogeneous plot. Type of leaf: 3rd or 4th leaf, from the apex of non-shaded branches. (As an alternative, collect a leaf with an axially located floral bud soon to be opened). Autumn.	N - 43-55(33-43) P - 2,3-2,7(1,2-2,1) K - 20-30(22-27) Ca - 9-25(12-16) Mg - 1,9-2,4(2,5-3,1) S - 3,2-4	B - 40-100 Cu - 10-15 Fe - 120-200 Mn - 40-250 Mo - 1,0-1,2 Zn - 25-60	[4]
Peach	100 leaves (4/tree) for each homogeneous plot. Type of leaf: recently matured and totally expanded.	N - 30-35 P - 1,4-2,5 K - 20-30 Ca - 18-27 Mg - 3-8 S - 1,5-3	B - 20-60 Cu - 5-16 Fe - 100-250 Mn - 40-160 Mo - Zn - 20-50	[4]
Grape	100 leaves (1/tree) for each homogeneous plot.	N - 30-35 P - 2,4-2,9	B - 45-53 Cu - 18-22	[6]

Culture	Plant organ; number of samples and period of sampling	Adequate dosages		References
		Macronutrients (g kg^{-1})	Micronutrients (mg kg^{-1})	
	Type of leaf: the youngest, recently matured, from branch apices.	K - 15-20 Ca - 13-18 Mg - 4,8-5,3 S - 3,3-3,8	Fe - 97-105 Mn - 67-73 Mo - Zn - 30-35	
	--- Cereals ---			
Corn	30 leaves/ha, of a homogenous plot showing female inflorescence (hair). Type of leaf: leaf obsta and below the corn ear	N - 27,5-32,5 P - 2,5-3,5 K - 17,5-22,5 Ca - 2,5-4 Mg - 2,5-4 S - 1,5-2	B - 15-20 Cu - 6-20 Fe - 50-250 Mn - 50-150 Mo - 0,15-0,2 Zn - 15-50	[8]
Sorghum	30 leaves/ha of a homogeneous plot at the start of tillering. Type of leaf: median	N - 13-15 P - 4,0-8,0 K - 25-30 Ca - 4-6 Mg - 4-6 S - 0,8-1	B – 20 Cu - 10 Fe - 200 Mn - 100 Mo - Zn – 20	[8]
	--- Forest species ---			
Eucalyptus	18 leaves/ha of a homogeneous plot, in Summer- Autumn. Type of leaf: recently matured primary branches in the superior third of the plant.	N - 14-16 P - 1-1,2 K - 10-12 Ca - 8-12 Mg - 4-5 S - 1,5-2	B- 40-50 Cu - 8-10 Fe - 150-200 Mn - 100-600 Mo - 0,5-1 Zn - 40-60	[8]
Pinus	18 leaves/ha of a homogeneous plot in Summer-Autumn. Type of leaf: Recently matured , primary	N - 12-13 P- 1,4-1,6 K - 10-11 Ca - 3-5 Mg - 1,5-2 S - 1,4-1,6	B - 20-30 Cu - 5-8 Fe - 50-100 Mn - 200-300 Mo - 0,1-0,3 Zn- 34-40	[8]
	--- Oilseeds ---			
Peanut	30 leaves/haof a homogeneous plot at the start of flowering. Type of leaf: 4th leaf of the main stalk from the basis (1ª = above the cotyledon air ebrabches).	N – 40 P - 2 K - 15 Ca - 20 Mg - 3 S - 2,5	B - 140-180 Cu - Fe - Mn - 110-440 Mo - 0,13-1,39 Zn -	[8]
Sunflower	30 leaves/ha of a homogeneous plot at the start of flowering.	N - 33-35 P - 4-7	B - 50-70 Cu - 30-50	[8]

Culture	Plant organ; number of samples and period of sampling	Adequate dosages		References
		Macronutrients (g kg^{-1})	Micronutrients (mg kg^{-1})	
	Type of leaf: leaves of the upper third.	K - 20-24 Ca -17-22 Mg - 9-11 S - 5-7	Fe - 150-300 Mn - 300-600 Mo - Zn - 70-140	
Soybean	30 leaves/ha of a homogeneous plot at the end of flowering. Type of leaf: first matured leaf from the branch end, excluding the petiole. General lythea 3rdleaf	N - 45-55 P - 2,6-5,0 K - 17-25 Ca - 4-2 Mg - 3-10 S - 2,5	B - 21-55 Cu - 10-30 Fe - 51-350 Mn - 21-100 Mo - Zn - 21-50	[8]
	-- Saccharine --			
Sugarcane (Plant)	20-30 leaves/ha of a homogeneous plot Type of leaf: leaf +3; leaf +1 = with the first ligula (=membranous outgrowth at the junction between the leaf blade and the sheath). Median third excluded the main rib	N - 19-21 P - 2-2,4 K - 11-13 Ca - 8-10 Mg - 2-3 S - 2,5-3	B - 15-50 Cu - 8-10 Fe - 200-500 Mn -100-250 Mo - 0,15-0,3 Zn - 25-30	[8]
Sugarcane (Ratoon)	20-30leaves/ha of a homogeneous plot, 4 month after sprouting. Type of leaf: leaf +3; leaf +1 = with first ligula (=membranous out growth at the junction between the leaf blade and the sheath). Median third excluded the main rib.	N - 20-22 P - 1,8-2 K - 13-15 Ca - 5-7 Mg - 2-2,5 S - 2,5-3	B - Cu - 8-10 Fe - 80-150 Mn - 50-125 Mo - Zn - 25-30	[8]
	-- Vegetable crops --			
Potato	30 leaves/ha of a homogeneous plot, in the middle of the cycle, 30-45 days after emergence. Type of leaf: Petiole of the 4^{ty} leaf from the tip.	N – 30 P - 3,5 K - 50 Ca - 20 Mg - 7,5 S - 3,5	B - 40-50 Cu - 5-8 Fe- 800-1000 Mn - Mo - Zn -	[8]
Onion	40 tip leaves/ha of a homogeneous plot at the middle of the cycle. Type of leaf: the highest one.	N – 40 P - 3 K - 40 Ca - 4 Mg - 4 S – 7	B - Cu - Fe - Mn - Mo - Zn -	[8]
Tomato	40 leaves/ha of a homogeneous plot in full flowering, or first ripe fruit;	N – 30 P - 3,5	B - 50-70 Cu - 10-15	[8]

Culture	Plant organ; number of samples and period of sampling	Adequate dosages		References
		Macronutrients (g kg⁻¹)	Micronutrients (mg kg⁻¹)	
	Type of leaf: 4th from the tip.	K - 40 Ca - 14-18 Mg - 4 S – 3	Fe - 500-700 Mn - 250-400 Mo - 0,3-0,5 Zn - 60-70	
	--- Stimulants ---			
Coffee	At least 30 days after the 2nd fertilizer portion or after one foliar spraying, in the pinhead phase, that is, before grain filling (December) sample the 3rd or 4th pair of leaves from the apex of productive branches, located in the plant median portion. Collect two pairs of leaves in both sides of the row in a total of 25 plants /homogeneous area sampled (100 leaves/ sample).	N - 29-32 P – 1,2-1,6 K - 18-22 Ca - 10-13 Mg - 3,1-4,5 S - 1,5-2	B - 40-80 Cu - 8-16 Fe - 70-180 Mn - 50-200 Mo - 0,1-0,2 Zn - 10-20	[9]
Cocoa tree	18 leaves/ha of a homogeneous plot in the Summer. Type of leaf: 3rd leaf from the tip, mature in plants half-shade.	N - 19-23 P - 1,5-1,8 K - 17-20 Ca - 9-12 Mg - 4-7 S - 1,7-2	B - 30-40 Cu - 10-15 Fe - 150-200 Mn - 150-200 Mo - 0,5-1,0 Zn - 50-70	[8]

Table 2. Procedures for leaf collection and ranges considered adequate of macro and micro nutrients contents in some cultures.

The table shows how important it is to follow the same recommendation (standard) for sampling and after the comparison of results. Collection mistakes that lead to wrong diagnostics and recommendations are common. It is emphasized that the main factors responsible for different nutritional levels in plants are:1- plant age; 2- organ analyzed; 3-type of plant (species, variety, graft/stock, crown);4- period of the year; 5- method of sample cleaning, extraction and quantification of nutrients; 6- water percentage in soil (for nutrients determined in the sap); 7- time of day (for nutrients determined in the sap);8- inadequate production of dry matter from the plant due to isolated or interative soil, climate, genotypic or human imperfections [1].

5. Preparation of vegetal material for analysis

In the laboratory the collected plant material is decontaminated (only the fresh non-dried material), dried, ground, the residual humidity is determined followed by weighing, nutrients quantification and results expressed. More details about the stages in plant analysis, including the determination of macro and micro nutrients is reported in [12] and [8].

5.1. Sample decontamination

According to [12], when only macro nutrients are to be determined in samples washing may be plain, to eliminate gross contaminations like dust. Just shaking the sample under tap water and rinsing with distilled water will be enough but the procedure must be fast to avoid the loss of soluble elements. Procedures are more elaborate when determination of microelements is contemplated. In this case the samples must be successively washed in tap water, dilute non -ionic detergent (0.1%, v/v), distilled water to remove detergent, 0.1M HCl, distilled water and finally deionized water. With samples highly soiled the battery of solvents must be changed as necessary. To avoid loss of soluble inorganic constituents the washing stages must not take more than 30 seconds.

Contaminations by pesticides and foliar fertilizers (especially when applied with surfactants in the spraying mixture) are difficult to remove by washing. Collection of samples in these cases must be carefully overseen.

5.2. Drying

Drying of samples must be as fast as possible to minimize biological and chemical alterations. After eliminating excess water, samples packed in paper bags are dried in 65 to 70ºC ovens fitted with devices for forced air circulation [12]. According to [1] temperatures must be higher, 70 to 80ºC to avoid putrefaction especially if samples are too close together. Samples should be kept in the oven till constant weight, which will be attained after 48 to 72 hours, depending on the vegetal material.

5.3. Grinding

Mills provided with stainless steel or plastic chambers are recommended to grind the vegetal material to reduce to a minimum contamination by micronutrients like Fe and Cu. Grinding is necessary to homogenize samples for analytical determinations and it must produce material that can be sieved through 1 to 20 mesh when using Wiley type mills. When alternating grinding of different samples mills are cleaned by brushing with 70% alcohol between procedures.

5.4. Chemical analysis: extraction and determination of nutrients

Chemical quantification of nutrients is the next step in the diagnosis of the nutritional status of a foliar sample. Several factors are involved in the choice among the different analytical methods available for this purpose. Some of them are: safety (hazard or toxicity), equipment available, type of element to be determined, precision and accuracy, period of time taken by analysis, limit of detection and cost [13].

In the laboratory the sample will be submitted to the following procedures: weighing, preparation of the extract and element determination (Figure 1).

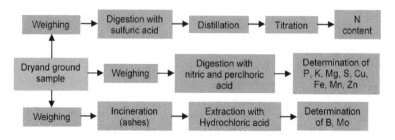

Figure 1. Simplified schematic procedure of foliar analysis to be conducted in a plant nutrition laboratory.

6. Interpretation of analytical results

Results are interpreted by comparing the concentration values of each element in the sample with the respective standard or a value considered optimal.

The foliar chemical analysis may be expressed in different methods, the most used are: 1- in the single variable methods only one of the elements is selected and the results are expressed by the deviation of the optimum percentage, the critical level and the sufficiency range; 2- the relation between the concentration values of nutrients is the basis for the double variable method named DRIS (Diagnosis and Recommendation Integrated System), or 3- the multivariable method named NCD (Nutritional Composition Diagnosis).

The flow chart in Figure 2 shows all the steps involved in foliar diagnosis starting with sampling up to the results obtained.

6.1. Sufficiency range

Most cultures do not have a single definite point for optimal production but a range of nutrient concentrations. Thus, it is adequate to recommend degrees of fertilization to keep nutrients slightly above the critical level, but included in the sufficiency range [14]. However, both have limitations the critical level by its precise character and the sufficiency range for lack of precision due to very wide limits.

The use of the sufficiency range is an attempt to extend a single optimal point into an optimal range and to make sure that at its highest level the culture is adequately supplied and at the lowest level it is so deficient that production will be negatively affected [1]. Generally, the sufficiency range corresponds to 90-100% of maximal production [15]. Also, the lowest limit of the sufficiency range will be the minimal critical point and superior limit the toxic critical point [1].

The ratio, foliar concentration and production is characterized by different ranges or zones (Figure 3), which should be discussed as detailed by [16].

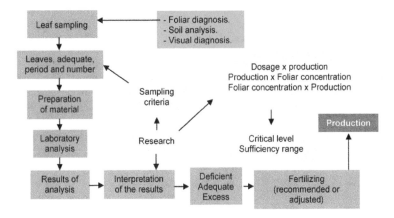

Figure 2. Flow chart for evaluation of the nutritional status of plants and its expansions according to the critical level and sufficiency range.

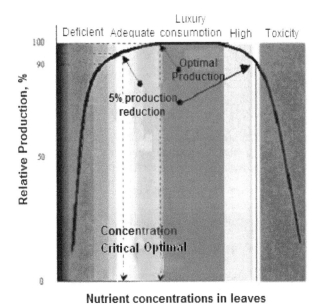

Figure 3. Relation of nutrient concentration and relative production.

1. In the deficient range or zone the symptoms are visible and occur in soils (or substrates) very deficient in an element due to insufficient dosages. In these conditions the response in production of dry matter is high, the element concentration is not increased and it may even be diluted. The nutrient dilution effect caused by organic matter formation is known as the Steembjerg effect. When the concentration of a plant nutrient is set in this range it is considered deficient.

2. In the transition range or zone, deficiency symptoms are not visible (disguised hunger) but there is a direct relation between nutrient foliar concentration of and production. When the nutrient concentration allows an average of 80 to 95% of the maximum production, this level corresponds to the critical level. The relation of nutrient concentrations and maximal production (100%) is seen in soils (or substrates) with slight deficiency and with lower responses in growth and production when the nutrient is applied. In these conditions the increases in foliar concentrations are proportional to growth and production, that is, greater absorption is compensated by increasing organic material. A nutrient concentration in this range, considered between the critical level and maximal production is interpreted as adequate.

3. In the luxury consumption range or zone, increasing element concentration does not increase production. This is observed in non-deficient soils receiving element dosages.

Although plant tissues show absorption of the increased nutrient concentrations this is not expressed in increased growth. Thus, the element concentration in this range, which corresponds to maximal or optimal production and it is below the toxicity critical point, is considered to be high.

4. The toxicity range or zone starts when increased nutrient concentrations significantly reduce production. Reductions of 5% up to 20% indicate toxic levels. The condition is observed in soils (or substrates) with excess nutrients receiving additional dosages that are absorbed as shown by increased tissue concentrations but expressed in decreased growth and/or imbalance in relation to other nutrients.

The critical level of deficiency is a factor largely employed in research and it corresponds to an optimal nutrient concentration. Below it the growth index (production or quality) is significantly decreased and above it, production represents poor economics.

After attaining maximal production, increased nutrient concentrations will not result in growth but in plant "luxury consumption". During this period nutrients accumulate in cell vacuoles and may be gradually liberated to supply eventual plant nutritional necessities. As already stated nutrient concentrations above the level of luxury consumption can lead to decreased production and characterize the toxicity range.

Interpretation of foliar nutrient concentrations based on the critical level and the sufficiency range is made directly by comparison with standard values. The plant nutritional status (deficiency, sufficiency, luxury consumption) is defined independently for each element by the range of values found for the sample. However, the plant mineral composition is the result of its adaptation to an environment under the action of several limiting factors. Lack of consideration of well-known and documented interactions between elements is severely criticized in these methods [17, 18].

6.2. Deviation from the percentage optimum

The deviation from the percentage optimum (DPO) is an improvement of the critical level method [19]. It evaluates each nutrient concentration in relation to the optimum value (median of the sufficiency range) by the expression: DPO= $[(Cx100/C_R)-100]$ where C is an element concentration in the sample dry matter and C_R it is the optimal concentration for the same conditions (culture, tissue analyzed, manipulation, plant development stage etc.). In the absence of the sufficiency range the critical level is taken as the optimum value.

This is a procedure not common in the literature but it permits the evaluation of the nutritional status of the plant and the arrangement of the elements as a function of the degree of deficiency. However, the limitation order is not representative because element interactions are not considered and the conventional table is still used.

6.3. Diagnosis and Recommendation Integrated System (DRIS)

DRIS is an alternative to the conventional method for the determination of the nutritional status of a plant [20]. It considers nutrient interactions in the diagnostic process, which is conducted by the combination of all the relations in the form of ratios [20] or products [21]. In this technique indexes, which express nutrient equilibrium in a plant or culture are calculated for each one, as a function of concentration ratios of each element and the total and compared in groups of two to other ratios considered standard or norms obtained in a population of highly productive plants.

Foliar diagnosis, in this method, aims to adjust fertilization, so far only recommended by soil fertility and culture productivity, by additional production gains and correction of deficiencies. It also makes possible the management other nutrient availabilities, possibly reducing them and permitting an equilibrated fertilization, in view of the culture nutritional necessities.

6.4. Diagnosis of nutritional composition (DNC)

The method relates nutrient concentrations in a multivariable form, as a function of ratios of each nutrient concentration and the geometrical mean of the nutritional composition of the sampled tissue [23]. The method is not widely used although it deals with relations between all elements analyzed.

DNC and DRIS are independent calibration methods, since use of double or multi variable methods minimizes non controlled effects of accumulated biomass, in contrast to the critical level, which needs calibration assays conducted in places and different years, and maintain control on other production factors (including other nutrients) and on a supply adequate to full plant development [24].

However, it is important to emphasize that all methods that interpret foliar analysis results are based on analysis of nutrient concentrations in plant dry matter. Thus, all procedures described in the previous topics (excluding biotic and abiotic factors that may interfere in the collection, preparation and analysis of sample and results) should be well conducted, since no analytical or interpretative method will correct mistakes in these steps.

7. Visual diagnosis

Visual plant nutrition diagnosis aims a detailed characterization of deficiency or toxicity symptoms in a plant-problem and compare them to standard patterns of deficiency or toxicity described in the literature.

To use this diagnosis it is necessary to make sure that the problem is caused by deficiency or excess of a nutrient, and not by pests and other diseases that may "masque" the problem by producing similar symptoms.

The symptoms caused by nutritional disorder generally have the following characteristics:

1. Dispersion- nutritional problems usually occur in the fields in a homogeneous form. In cases of pest/diseases the occurrence may be limited to isolated plants or dense growth. Nutritional deficiencies rarely appear only in some plants.

2. Simetry- nutritional disorders usually occur symmetrically in leaves while phytopathogenic or insect injuries provoke asymmetrical symptoms with the exception of the ones caused by viruses, which translocate though the whole plant and may produce foliar symptoms similar to nutritional deficiency.

3. Gradient- in a plant or branch the symptoms appear in a gradient, becoming more severe going from old to young leaves or in reverse, according to the element mobility in the plant.

In visual diagnosis symptoms of deficiency/excess may vary in cultures. Generally, deficiency signs start in older leaves for the easily distributed elements and in new leaves and shoots for elements of lower redistribution. The signs may be visualized in roots, like in conditions of Al toxicity, which induces ill-formed roots, thick and short. Visual symptoms of nutritional deficiency may be grouped in six categories: a) reduced growth; b) uniform chlorosis or leaf spots; c) interrib chlorosis; d) necrosis; e) red color; f) deformities.

The visual diagnosis method allows for fast identification of deficiencies or excesses with consequent correction of fertilization. However, it is a limited method criticized by some authors as described in [17].

• In the field the plant may suffer from interfering agents (pests and pathogens) that mimetize nutritional deficiency symptoms, as already stated.

- Deficiency symptoms may be different from the ones described in the literature or specialized publications. For example, symptoms may be light instead of the severe ones described.

- Element deficiency signs may be different according to element and culture. Zn deficiency in fruit trees is expressed by smaller leaves and in corn cultures, new leaves are bleached.

- Deficiency symptoms may be similar for different nutrients.

- Certain deficiencies may reduce production without plant symptoms.

- Deficiencies of two or more nutrients prevents identification.

- Excess of one nutrient may be mistakenly taken as the deficiency of another one.

- Adequate visual diagnosis must be conducted by technicians with significant experience in cultures of the region.

- Visual diagnosis does not quantify neither the deficiency level nor the excessive one.

Furthermore, when the nutritional disorder is acute and visual symptoms of deficiency or excess are obvious and able to be differentiated a significant part of production (around 40-50%) may have been already compromised by a series of irreversible injuries to the physiology of the plant. Thus, visual diagnosis should not be used as a rule but only as complement.

8. Other methods

Foliar diagnosis is a direct evaluation method that utilizes nutrient concentrations in plant tissues as an indicator of nutritional status. However, indirect methods exist and are useful. When a deficient nutrient is part of an organic component or activates an enzymic activity this can be indirectly expressed. For example N deficiency may be shown by low chlorophyll levels or low activity of nitrate reductase. A description of biochemical tests that may be employed to evaluate plant nutritional status has been reported in [8]. For N, reductase and glutamine synthetase activity, amide N and asparagine; for P, fructose-1,6-diphosphate and photosynthesis ; phosphatase activity; for K,amide concentrations; free amino-acids; for Mn, peroxidases and a/b chlorophyll ratios; for B, ATP-ase activity; for Zn, ribonuclease, carbonic anhydrase, arginine concentration. In the case of P other studies indicate that P_i in vacuole cells may indicate the nutritional status of the plant [25, 26]. These are additional tools to evaluate plant nutrition, which are not commonly used because some of the tests require special methods of sampling, storage and complex analytical procedures and costly equipment. Other methods, specifically for N, evaluate the index of green color by a portable device called chlorophyll meter. This index is strongly correlated to the chlorophyll concentration in leaves and N nutritional status of the plant.

9. Final considerations

Plant analysis is a fundamental tool for nutritional diagnoses in cultures. The technique permits control of the nutritional equilibrium in cultures, reduction of costs and avoids environmental impact though rational use of fertilizers and consequent gains in production and profit. The main difficulties in the procedure, leaf sampling and interpretation of analysis results but these are improving as time goes by, becoming safer, economical fast and precise. The non-standardized sampling techniques diverge among author preferences but are intensely researched and improved by recommendations in comparative studies between samples and standard. Results interpretation is mainly by the critical level and sufficiency range. Alternative methods, like DRIS and DNC have been proposed but their use is still incipient.

Efficient fertilization calls for equal consideration and care to all phases of the process as plant sampling in the fields, laboratory analysis but mostly it should conducted by competent and experienced professionals In addition it is recommended the integrated utilization of techniques, that is, chemical analysis must be complemented by visual diagnosis so that, fertilization is efficient, economically profitable but safe to the environment.

Author details

Renato de Mello Prado* and Gustavo Caione*

*Address all correspondence to: rmprado@fcav.unesp.br

*Address all correspondence to: gustavocaione@agronomo.eng.br

UNESP (Universidade Estadual Paulista), Jaboticabal, SP, Brazil

References

[1] Fontes PCR. Diagnóstico do estado nutricional das plantas. 1º.ed. Viçosa, UFV; 2001. 122p.

[2] Gan S., Amasani, RM. Making sense of senescence. Plant Physiology 1997; 113: 313-319.

[3] Veloso CAC., Araújo SMB., Viégas, IJM., Oliveira RF. Amostragem de Plantas para Análise Química. Empresa Brasileira de Pesquisa Agropecuária – EMBRAPA; 2004. Comunicado Técnico, 121, Belém.

[4] Raij B.van, Silva NM., Bataglia OC., Quaggio J.A., Hiroce R., Cantarella H., Bellinazzi Júnior R., Dechen AR., Trani PE. Recomendações de adubação e calagem para o esta-

do de São Paulo. Campinas, Instituto Agronômico de Campinas e SAAESP; 1985. 107p. (Boletim técnico, 100).

[5] IFA. World fertilizer use manual. Paris: International Fertilizer Industry Association; 1992. p.283-284.

[6] Quaggio JA., Raij B.van, Piza Jr. CT. Frutíferas. In: Raij, B. Van; Cantarella, H.; Quaggio JA., Furlani AMC. (Eds.) Recomendações de adubação e calagem para o Estado de São Paulo. 2.ed. Campinas: Instituto Agronômico/Fundação IAC; 1996. p.121-153.

[7] Natale W., Coutinho ELM., Boaretto AE., Pereira FM., Modenese SH. Goiabeira: calagem e adubação. Jaboticabal: FUNEP; 1996. 22p.

[8] Malavolta E., Vitti GC., Oliveira SA. Avaliação do estado Nutricional das plantas: princípios e aplicações. 2º ed. Piracicaba: Associação Brasileira para Pesquisa da Potassa e do Fosfato; 1997. 319p.

[9] Guimarães PTG., Garcia AWR., Alvarez V HV., Prezotti LC., Viana AS., Miguel AE., Malavolta E., Corrêa JB., Lopes AS., Nogueira FD., Monteiro AVC., Oliveira JA. Cafeeiro. In: Ribeiro AC., Guimarães PTG., Alvares V HV. (Eds). Recomendação para uso de corretivos e fertilizantes em Minas Gerais (5º Aproximação). Viçosa; 1999. p. 289-305.

[10] Borges AL., Raij B van., Magalhães AF., Bernardi AC. Nutrição e adubação da bananeira irrigada. Cruz das Almas: Embrapa Mandioca e Fruticultura; 2002. 8p. (Embrapa-CNPMF. Circular Técnica, 48).

[11] Silva JTA., Borges AL., Dias MSC., Costa EL., Prudêncio JM. Diagnóstico nutricional da bananeira 'Prata-Anã' para o Norte de Minas Gerais. Belo Horizonte: Epamig; 2002.16p. (Boletim Técnico, 70).

[12] Bataglia OC., Furlani AMC., Teixeira JPF., Furlani PR., Gallo JR. Métodos de análise química de plantas. Campinas, Instituto Agronômico; 1983. 48p.

[13] Nogueira ARA., Matos AO., Carmo CAFS., Silva DJ., Monteiro FL., Souza GB., Pita GVE., Carlos GM., Oliveira H., Comastri Filho JA., Miyazawa M., Oliveira Neto WT. Tecido vegetal. In: Nogueira A., Souza GB. (Eds.). Manual de laboratórios: solo, água, nutrição vegetal, nutrição animal e alimentos. São Carlos: Embrapa Pecuária Sudeste; 2005. p.145-199.

[14] Bataglia OC., Dechen AR., Santos WR. Diagnose visual e análise de plantas. In: Reunião Brasileira de Fertilidade do Solo d Nutrição de Plantas, 20, Anais... Piracicaba, SBCS; 1992. p.369-404.

[15] Jones Jr. JB. Interpretation of plant analysis for several agronomic crops. In: Walsh LM., Beaton JD. (Eds). Soil testing and plant analysis. Part 2. Madison: SSSA; 1967. p. 49-58.

[16] Prado RM. Nutrição de plantas. São Paulo: Editora UNESP; 2008. 407p.

[17] Marschner H. Mineral nutritionof higher plants. New York: Academic Press; 1986. 674p.

[18] Malavolta, E. Manual de nutrição de plantas. Piracicaba: Ceres; 2006. 631p.

[19] Monatanes L., Heras L., Sanz M. Desviación del optimo porcentual (DOP): nuevo índice para la interpretación del análisis vegetal. Anales de Aula Dei 1991; 20 (3-4): 93-107.

[20] Beaufils ER. Diagnosis and recommendation integrated system (DRIS): a general scheme of experimentation and calibration based on principles developed from research in plant nutrition. Pietermaritzburg: University of Natal; 1973. 132p.

[21] Walworth JL., Sumer RE. The diagnosis and recommendation integrated system (DRIS). Advances Soil Science 1987; 6: 149-188.

[22] Serra AP., Marchetti ME., Vitorino ACT., Novelino JO., Camacho MA. Determinação de faixas normais de nutrientes no algodoeiro pelos métodos CHM, CND e DRIS. Revista Brasileira de Ciência do Solo 2010; 34: 105-113.

[23] Parent LE., Dafir M. A theoretical concept of compositional nutrient diagnosis. Journal of the American Society for Horticultura Science 1992; 117: 239-242.

[24] Parent LE., Natale W. CND: Vantagens e benefícios para culturas de alta produtividade. In: Prado RM., Rozane DE., Vale DW., Correia MAR., Souza HA. (Eds.). Nutrição de Plantas: Diagnose foliar em grandes culturas. Jaboticabal, Faculdade de Ciências Agrárias e Veterinárias, Universidade Estadual Paulista, FUNDENESP; 2008. p.105-114.

[25] Bollons HM., Barraclough PB. Assessing the phosphorus status of winter wheat crops: inorganic orthophosphate in whole shoots. Journal of Agricultural Science 1999; 133: 285-295.

[26] Bollons HM., Barraclough PB. Inorganic orthophosphate for diagnosing the phosphorus status of wheat plants. Journal of Plant Nutrition 1997; 20 (6): 641-655.

Chapter 5

Nutrient Balance as Paradigm of Soil and Plant Chemometrics

S.É. Parent, L.E. Parent, D.E. Rozanne,
A. Hernandes and W. Natale

Additional information is available at the end of the chapter

1. Introduction

Soil fertility studies aim to integrate the basic principles of biology, chemistry, and physics, but generally lead to separate interpretations of soil and plants data [1]. Paradoxically, J.B. Boussingeault warned as far as in the 1830s that the balance between nutrients in soil-plant systems was more important than nutrient concentrations taken in isolation [2]. Indeed, the biogeochemical cycles of elements that regulate the dynamics of agroecosystems [3] do not operate independently [4]. However, raw concentrations of individual elements or their log transformation are commonly used to conduct statistical analyses on plant nutrients [5, 6,7], soil fertility indices [8] and C mineralization data [9, 10]. Researchers thus proposed several ratios and stoichiometric rules to relate system's components to each other when monitoring mineralization and immobilization of organic C, N, P and S in soils [4, 11], cations interacting on soil cation exchange capacity [12], nutrient interactions in plants [13, 14, 15, 16] and carbon uptake by plants [17, 18].

Different approaches have been elaborated to describe nutrient balances in soils. The nutrient intensity and balance concept (NIBC) computes ionic balances in soil water extracts [19, 20]. The basic cation saturation ratio (BCSR) concept hypothesizes that cations and acidity exchanging on soil cation exchange capacity (CEC) can be optimized for crop growth [12]. However, the BCSR has been criticized for its elusive definition of 'ideal' cationic ratios [21, 22]. In plant nutrition, [23] were the first to represent geometrically interactions between nutrients by a ternary diagram where one nutrient can be computed by difference between 100% and the sum of the other two. As a result, there are two degrees of freedom in a ternary diagram. One may also derive three dual ratios from K, Ca, and Mg but only two ratios are linearly independent because, for example, K/Mg can be computed from K/Ca × Ca/Mg and is thus redundant. Therefore, a ratio approach conveys D-1 degrees of freedom or linearly independent balances for a D-part composition [24].

In contrast, there are $D\times(D-1)/2$ dual ratios such as the K/Mg ratio and $D\times(D-1)^2/2$ two-component amalgamated ratios such as the K/(Ca+Mg) ratio that can be derived from a D-part composition. Most information on dual and two-component amalgamated ratios is thus redundant and the dataset is artificially inflated. In Figure 1, the number of (a) dual and (b) two-component amalgamated ratios is plotted against the number of components. With 10 components, one may compute up to 45 dual and 405 two-component amalgamated ratios, hence generating a "redundancy bubble" that inflates exponentially above D. [25] elaborated the Diagnosis and Recommendation Integrated System (DRIS) to synthesize the $D\times(D-1)/2$ dual ratios into D nutrient indices adding up to zero; therefore, there is still one redundant index closing the system to zero and computable from other indices. Applying Ockham's razor law of parsimony to compositional data, nine degrees of freedom suffice to fully describe a 10-part composition without bias [24].

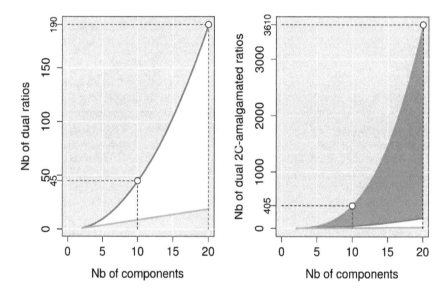

Figure 1. Number of (a) dual and (b) two-component amalgamated ratios illustrating the redundancy bubble.

To solve problems related to nutrient diagnosis in soil and plant sciences, one must first recognize that soil and plant analytical data are most often compositional, i.e. strictly positive data (concentrations, proportions) related to each other and bounded to some whole [26]. Compositional data have special numerical properties that may lead to wrong inferences if not transformed properly. Log-ratio transformations have been developed to avoid numerical biases [26, 29, 30, 31]. The balance concept presented in this chapter is based on log ratios or contrasts. Balances are computed rather simply from compositions using the isometric log-ratio (*ilr*) transformation developed by [27]. In the literature, the nutrient balance often refers to a nutrient budget that measures the depletion or accumulation of a given nutrient in soils [28], implying exchange between compartments of some whole. In this chapter, nutrient balance is defined as dual or multiple log ratios between nutrients, implying balance between components of the same whole.

The aim of this chapter is to introduce the reader to the balance concept as applied to soil fertility studies. The first section of this chapter presents the theory common to the three subsequent subjects, which are cationic balance in soils, plant nutrient signatures and mineralization of organic residues. It is suggested that the reader gets familiar with the theory before browsing through the subject of interest.

2. Theory of CoDa

Because a change in any proportion of a whole reverberates on at least one other proportion, proportions of components of a closed sum (100%) are interdependent. Therefore, a compositional vector is intrinsically multivariate: its components cannot be analyzed and interpreted without relating them to each other [32,33]. Compositional data (CoDa) induce numerical biases, such as self-redundancy (one component is computable by difference between the constrained sum of the whole and the sum of other components), non-normal distribution (the Gaussian curve may range below 0 or beyond 100% which is conceptually meaningless) and scale dependency (correlations depend on measurement scale). Redundancy can be controlled by carefully removing the extra degree of freedom in the *D*-part composition. Scale dependency is controlled by ratioing components after setting the same scale (e.g. fresh mass, dry mass or organic mass basis) or unit of measurement (e.g. mg kg^{-1}, g dm^{-3}, cmol$_c$ kg^{-1}, etc.) across components. Compositional datasets constrained to a closed space between 0 and 100% are amenable to normality tests after projecting them into a real space using log-ratio transformations.

One of the log ratio transformations is the centered log ratio (*clr*) developed by [26]. The *clr* is a log ratio contrast between the concentration of any nutrient and the geometric mean across the compositional vector. [34] used the *clr* to convert DRIS into Compositional Nutrient Diagnosis (CND-*clr*), hence correcting inherent biases generated by DRIS. [35] and [36] modeled the time change of ion activities in soils and nutrient solutions using *clr*. However, because *clr* generates a singular matrix (the *clr* variates sum up to 0), one *clr* value should be removed (e.g. that of the filling value) in multivariate analysis. In addition, outliers may affect considerably log ratios [32]. The diagnostic power of CND-*clr* is decreased by large variations in nutrient levels (e.g. Cu, Zn, Mn contamination by fungicides) that affect the geometric means across concentrations. Nevertheless, the *clr* transformation is useful to conduct exploratory analyses on compositional data [37].

The additive log ratio or *alr* [26] computed as $\ln(x/x_D)$ is the ratio between any component x and a reference component x_D. [17] used nitrogen as reference component (N=100%) to produce a stoichiometric N:P:K:Ca:Mg rule for adjusting nutrient needs of tree seedlings. If a tissue contains 2.50% N and 0.15% P, the Redfield N/P ratio [38] is 16.7 and the corresponding *alr* [P/N] value is ln(0.15/2.50) = -2.81. Other stoichiometric rules have been proposed such the C:N:P:S rule for humus formation [4]. There are *D*-1 *alr* variables in a *D*-part composition because one component is sacrificed as denominator. The *alrs* are oblique to each other and are thus difficult to rectify and interpret [24]. Orthogonal balances are log ratio contrasts between geometric means of two groups of components that are multiplied by orthogonal coefficients to gain orthogonality [27]. Orthonormal balances are called 'isometric log ratios' coordinates or *ilr* [27] and are illustrated by a mobile and its fulcrums (CoDa dendogram) [37]. Balances are encoded in a device called sequential binary partition that orderly allocates components to balance numerator and denominator or +/- sides of a contrast. The *ilr* of groups of components is a thus rectified ratio between their geometric means. Balances avoid matrix singularity and redundancy: there are *D*-1 independent balances in a *D*-part composition. The orthonormal balance concept was found to be the most appropriate technique in the multivariate [29] and multiple regression [39] analyses in geochemistry [40], plant nutrition [34, 35, 36, 41, 42], the P cycle [43], and soil quality [44, 45].

2.1. From CoDa to sound balances

The sample space of a compositional vector defined by S^D is a strictly positive vector of D nutrients adding up to some constant κ. The closure operation, $\mathcal{C}$, computes the constant sum assignment as follows :

$$S^D = \mathcal{C}(c_1, c_2, \ldots, c_D) = \left[\frac{c_1 \kappa}{\sum\limits_{i=1}^{D} c_i}, \frac{c_2 \kappa}{\sum\limits_{i=1}^{D} c_i}, \ldots, \frac{c_D \kappa}{\sum\limits_{i=1}^{D} c_i} \right] \tag{1}$$

Where $\sum\limits_{i=1}^{D} c_i$ closes the sum of components to some whole such as 1, 100%, 1000 g kg^{-1}, which allows computing a filling value to the unit of measurement. In other cases where the data do not add up to the measurement unit such as mg dm^{-3} or mg L^{-1}, the measurement unit just cancels out when components are ratioed.

In general, raw or log-transformed concentration data are analyzed statistically without any *a priori* arrangement of the data. The analyst not only processes such data through a numerically biased procedure, but also relies on a cognitively unstructured path that returns unstructured results that are barely interpretable (Figure 2).

Fortunately, recent progress in compositional data analysis provides means to elaborate structured pathways and interpret results coherently [27]. Indeed, the *ilr* technique transforms a *D*-part composition into *D*-1 pre-defined orthogonal balances of parts projected into a real Euclidean space [24]. Orthogonality is a special case of linear independence where vectors fall perfectly at right angle to each other [46]. The balances can thus be analyzed as additive (undistorted) variables in the Euclidean space, hence without bias. The log ratio of X/Y is also called a log contrast between X and Y because log(X/Y) = log(X) – log(Y). A log ratio can scan the real space (±∞) because ratios may range from large numbers (positive log values) to small fractions (negative log values).

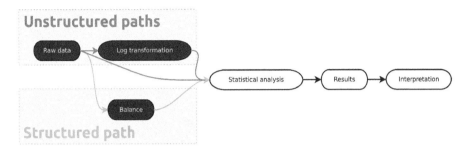

Figure 2. Unstructured or knowledge-based (structured) pathways of compositional data analysis lead to numerically biased and unbiased interpretations, respectively.

Balances can be illustrated by a CoDa dendrogram [37] where components or groups of components are balanced by analogy to a mobile and its fulcrums (Figure 3). Each part has its own weight and the balance between parts or groups of parts are the fulcrums (boxplots) equilibrating the system and computed as *ilr*. It can be shown that a relative increase in Ca concentration will change the [Ca | Mg] balance and [N,P,K | Ca, Mg] balances without affecting the ([N,P | K] and [N | P]. Transforming compositions to functional balances does not only create orthogonal real variables amenable to linear statistics; it also creates new variables whose interpretation is also of interest. Thus the interpretation of relationships between nutrients depends on how balances are conceived using the best science and management options. For example, another balance setup could be defined as [N,P | K, Ca, Mg], [N | P], [K | Ca, Mg] and [Ca | Mg].

A CoDa dendrogram (e.g. Figure 3) is interpreted as follows:

• Each fulcrum represents a balance. There are 4 balances for 5 components in Figure 3.

• If the fulcrum lies in the center of the horizontal bar, the balance is null. If it lies on the left side of the center, the mean balance is negative and left-side components occupy a larger proportion in the simplex. A fulcrum on the right side indicates a positive balance.

• Rectangles located on fulcrums are boxplots.

• The length of vertical bars represent the proportion of total variance

Nested balances are encoded in an *ad hoc* sequential binary partition (SBP) that nurtures the ties between groups of components. A SBP is a $(D-1){\times}D$ matrix, where parts labelled "+1" (group numerator) are balanced with parts labelled "-1" (group denominator) in each ordered row. A part labelled "0" is excluded. The composition is partitioned sequentially at every ordered row into 2 contrasts until (+1) and (-1) subcompositions each contain a single part. The analyst can use exploratory analysis [37] or refer to current theory and expert knowledge to design the balance scheme. The CoDa dendrogram in Figure 3 is formalized by the SBP in Table 1.

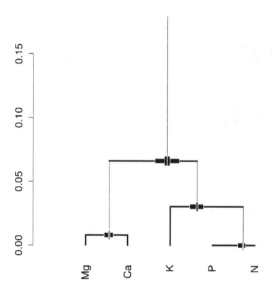

Figure 3. Balances between N, P, K, Ca, and Mg (five weight variables) are illustrated by a mobile and its fulcrums (four balance variables) where N, P, and K are contrasted with Ca and Mg, N and P with K, N with P, and Ca with Mg.

Binary partiton	Balance between groups of components					r	s	*ilr computation*
	N	P	K	Ca	Mg			
[N,P,K | Ca,Mg]	+1	+1	+1	-1	-1	3	2	$\sqrt{\frac{3\times2}{3+2}}ln\left(\frac{(N\times P\times K)^{1/3}}{(Ca\times Mg)^{1/2}}\right)$
[N,P | K]	+1	+1	-1	0	0	2	1	$\sqrt{\frac{2\times1}{2+1}}ln\left(\frac{(N\times P)^{1/2}}{K}\right)$
[N | P]	+1	-1	0	0	0	1	1	$\sqrt{\frac{1\times1}{1+1}}ln\left(\frac{N}{P}\right)$
[Ca | Mg]	0	0	0	+1	-1	1	1	$\sqrt{\frac{1\times1}{1+1}}ln\left(\frac{Ca}{Mg}\right)$

Table 1. Sequential binary partition defining macronutrient balances.

In Table 1, the sequential binary partition of nutrients encodes the balances between two geometric means across the + components at numerator and the – components at denominator. The orthogonal coefficient of a log contrast is computed from the number of + and – components in each binary partition. The balances between two subcompositions are orthogonal log ratio contrasts between geometric means of the "+1" and "-1" groups. The jth ilr coordinate is computed as follows [24]:

$$ilr_j = \sqrt{\frac{rs}{r+s}}ln\frac{g(c_+)}{g(c_-)} \text{ , with } j = \left[1, 2, \dots, D-1\right] \quad (2)$$

Where ilr_j is the jth isometric log-ratio; $g(c_+)$ is geometric mean of components in group "+1", c_+; and $g(c_-)$ is the geometric mean of components in group "-1", c_-. Because dual ratios are nested into $g(c_+)$ and $g(c_-)$, the balances avoid generating redundant ratios. The orthogonal coefficient is computed as $\sqrt{rs/(r+s)}$ [27]. For example, a Redfield N/P ratio of 16.7 is converted into ilr as $\sqrt{1x1/(1+1)}\ln(16.7)=2.02$. The ilr technique is thus not only mathematically elegant, but is also conceptually meaningful.

2.2. Dissimilarity between compositions

As a result of orthogonality, the Aitchison distance ($\mathcal{A}$) between any two compositions is computed as a Euclidean distance across the selected ilr coordinates as follows [47]:

$$\mathcal{A} = \sqrt{\sum_{j=1}^{D-1} \left(ilr_j - ilr_j^*\right)^2} \tag{3}$$

Where ilr_j is the jth ilr of a given composition and ilr_j^* is the corresponding ilr for the reference composition. Selecting alternative SBPs to test and interpret other balances in the system under study just rotates the orthogonal axes of the ilr coordinates without affecting $\mathcal{A}$. The Aitchison distances computed across ilr or clr values are identical [24]. [34] rectified DRIS to fit into clr. As computed from dual ratios and nutrient indices [13] and using the same reference population as reference for computing the Aitchison distance, the DRIS nutrient imbalance index appeared to be slightly distorted and noisy (Figure 4). Tissue analyses in Figure 4 were obtained from a survey across guava (*Psidium guajava*) orchards in the state of São Paulo, Brazil. Noise and distortion between results observed in Figure 4 is attributable to numerical biases in DRIS results.

On the other hand, the Euclidean distance ($\mathcal{E}$) based on log transformations is biased by the difference between the geometric means times the number of parts as follows [48]:

$$\mathcal{E}^2(\ln(x),\ \ln(y)) = \mathcal{A}^2 + D\left(\ln\frac{g(x)}{g(y)}\right)^2 \geq \mathcal{A}^2 \tag{4}$$

In plant nutrition studies [49], the Mahalanobis distance ($\mathcal{M}$) may be preferred to the Euclidean distance because the former takes into account the covariance structure of the data [29] (as illustrated in Figure 5) and has χ^2 distribution [50,51]. The M^2 is computed as follows:

$$\mathcal{M}^2 = (x - \bar{x})^T \times COV^{-1}(x - \bar{x}) \tag{5}$$

Where $\bar{x}$ is the mean and COV is the covariance matrix. Both $\mathcal{A}$ and $\mathcal{M}$ computed across log-transformed data are higher than their counterparts computed across balances, indicating systematic upper bias using natural log compared to ilr transformations (Figure 6). Tissue analyses in Figure 6 were obtained from the same guava orchard survey as above.

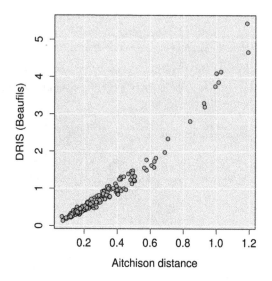

Figure 4. Distance from a reference composition computed using DRIS versus the Aitchison distance.

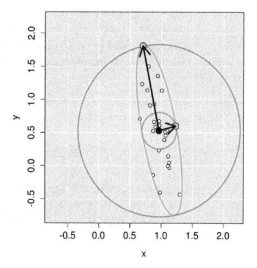

Figure 5. The Euclidean distance is circular while the Mahalanobis distance (*M*) is elliptical. The blue ellipse represents a line of equidistant points in terms of *M* that scales data to the variance in each direction. The *M* between green points and the center are equal. However, the Euclidean distance between each green point and the center is different, as shown by the Euclidean equidistance pink circles.

2.3. Cate-Nelson analysis

The Cate-Nelson procedure was developed as a graphical technique to partition percentage yield (yield in control divided by maximum yield with added nutrient) versus soil test [52]. The scatter diagram is subdivided into four quadrants to determine a critical test level and a critical percentage yield by maximizing the number of points in the + quadrants. This technique is analog to binary classification tests widely used in medical sciences [53] where data each quadrant are interpreted as true positive (correctly diagnosed as sick), false positive (incorrectly diagnosed as sick), true negative (correctly diagnosed as healthy) and false negative (incorrectly diagnosed as healthy). Applied to soil fertility studies, we can define four classes as follows:

- True positive (TP: nutrient imbalance): imbalanced crop (low yield) correctly diagnosed as imbalanced (above critical index).

- False positive (FP: type I error): balanced crop (high yield) incorrectly identified as imbalanced (above critical index). FP points indicate luxury consumption of nutrients.

- True negative (TN: nutrient balance): balanced crop (high yield) correctly diagnosed as balanced (below critical index).

- False negative (FN: type II error): imbalanced crop (low yield) incorrectly identified as balanced (below critical index). FN points show impacts of other limiting factors.

The performance of the test is measured by four indices:

- Sensitivity: probability that a low yield is imbalanced as TP/(TP+FN)

- Specificity: probability that a high yield is balanced as TN/(TN+FP)

- Positive predictive value (PPV): probability that an imbalance diagnosis returns low yield as TP/(TP+FP)

- Negative predictive value (NPV): probability that a balance diagnosis returns high yield as TN/(TN+FN)

The performance of the binary classification test is higher when the four indexes get closer to unity. However, the maximization of the four indexes may not be the most appropriate procedure. Indeed, agronomists are more interested in high PPV than in high specificity.

Using the Cate-Nelson graphical procedure, the TN specimens are selected as reference population after removing outliers. If the number of points is too large, yields are arranged in an ascending order and a two-group partition is computed. The sums of squares between two consecutive groups of observations are iterated as follows:

$$Class\ sum\ of\ squares = \frac{\left(\sum_{j=1}^{n_1} Y_{1j}\right)^2}{n_1} + \frac{\left(\sum_{j=1-n_1}^{n} Y_{2j}\right)^2}{n_2} - \frac{\left(\sum_{j=1}^{n} Y_j\right)^2}{n} \qquad (6)$$

Where Y_{1j} is class 1 yields starting with the two lowest soil indices; the remaining yields are in class 2 or Y_{2j}; and n_1, n_2 and n are the numbers of observation in class 1, class 2 and across classes, respectively. The last member of the equation is the correction factor. The starting values for maximization of the sums of squares across $\mathcal{A}$ or $\mathcal{M}$ could be the *ilr* means of the upper 20 top specimens [54]. Due to yield variations between production years, the upper quartile of higher yield standardized by year of production is an additional option. Because the iterative procedure is very sensitive to extreme values, an *a posteriori* visual adjustment may be necessary to maximize the number of points in opposite quadrants.

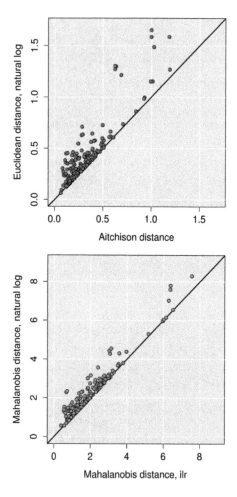

Figure 6. Relationships between the Euclidean, Aitchison and Mahalanobis distances.

2.4. Statistics

In this chapter, statistics computed across compositional data were performed in the R statistical environment [55]. Compositional data analysis was conducted using the R "compositions" package [56]. Data distribution was tested using the Anderson-Darling normality test [57] in the "nortest" package [58]. Multivariate outliers were removed using $\mathcal{M}$ computed in the R "mvoutlier" package [59]. Linear discriminant analysis (LDA) was used as a statistical ordination technique that allows computing linear combinations of variables that best discriminate groups. Multiple regression analysis was conducted using ilr [39] and compared to raw data. After completing the statistical analysis, the balances could be back-transformed to the familiar concentration units using the D-1 ilr values and the sum constraint.

3. Cationic balances in tropical soils

3.1. Sequential binary partition

The percentage base saturation is the proportion of soil cation exchange capacity (CEC) occupied by a given cation. The soil compositional vector is defined as follows [12]:

$$S^4 = \mathcal{C}(K, \ Ca, \ Mg, \ H + Al) \tag{7}$$

As illustrated in Figure 7, the first contrast, [K | Ca, Mg, H+Al], balances the K against divalent cations and acidity to enable adjusting the K fertilization to soil basic acid-base conditions as modified by liming.

The second contrast [Ca, Mg | H+Al] is the acid-base contrast for determining lime requirements while the [Ca | Mg] balance reflects the Ca:Mg ratio in soils adjustable by the liming materials. Alternative SBPs could also be elaborated such as [K, Ca, Mg | (H+Al)], [K | Ca, Mg] and [Ca | Mg] balances that reflects the BCSR model of [12]. The selected sequential binary partition for cationic balances is presented in Table 2.

For example, if a soil contains 2.9 mmol$_c$ K dm^{-3}, 20 mmol$_c$ Ca dm^{-3}, 5 mmol$_c$ Mg dm^{-3}, and 23 mmol$_c$ H+Al dm^{-3}. Cationic balances are computed as follows:

$$\left(1\right)[K \mid Ca, Mg, H + Al] = \sqrt{\tfrac{1 \times 3}{1 + 3}} \ln\left(\tfrac{2.9}{\sqrt{20 \times 5, \ x23}}\right) = -1.312;$$

$$\left(2\right)[K \mid Ca, Mg] = \sqrt{\tfrac{1 \times 2}{1 + 2}} \ln\left(\tfrac{2.9}{\sqrt{20 \times 5}}\right) = -1.011; \text{ and}$$

$$\left(3\right)[Ca \mid Mg] = \sqrt{\tfrac{1 \times 1}{1 + 1}} \ln\left(\tfrac{20}{5}\right) = 0.980.$$

Note that the K fertilization would depend on soil acidity as well as levels of exchangeable Ca and Mg in the soil. We thus expect the K index and the K balance to be similarly related to fruit yield if the *ceteris paribus* assumption applies to exchangeable Ca, Mg, and acidity in this soil-plant system.

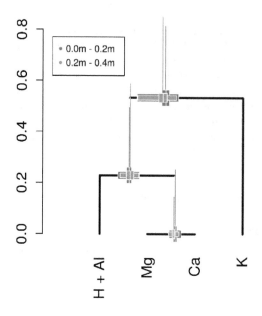

Figure 7. The cationic balances in soils can be designed to facilitate K fertilization and lime management. Each balance between two groups of ions is computed as isometric log ratio.

Partition	Cationic balances				r	s	ilr formulation
	K	Ca	Mg	H+Al			
1	1	-1	-1	-1	3	1	$\sqrt{\frac{1\times3}{1+3}}\ln\left(\frac{K}{\sqrt[3]{CaxMg,\,x(H+Al)}}\right)$
2	1	-1	-1	0	1	2	$\sqrt{\frac{1\times2}{1+2}}\ln\left(\frac{K}{\sqrt{CaxMg}}\right)$
3	0	1	-1	0	1	1	$\sqrt{\frac{1\times1}{1+1}}\ln\left(\frac{Ca}{Mg}\right)$

Table 2. Sequential binary partition of soil cationic data

3.2. Datasets

Changes in soil cationic balances were monitored in N and K fertilizer trials established on an epieutrophic and endodystrophic soil (Red-Yellow Oxisol) [60] at São Carlos (São Paulo, Brazil). One year old plants of 'Paluma' guava (*Psidium guajava*) were planted. The experiment lasted 3 yr. The N treatments in the 1st year were 0, 30, 60, 120, 180, 240 and 300 g N tree[-1] supplemented with 52 g P tree[-1] and 52 g K tree[-1]. The initial N rates were doubled and tripled in the 2nd and 3rd years, respectively. The initial P and K doses were doubled the 2nd year. The 3rd year, rates were 240 g P_2O_5 tree[-1] and 360 g K_2O tree[-1]. Fertilizers were ammonium nitrate (34% N), simple superphosphate (8.7% P) and potassium chloride (50% K). In the K trial, K was added as KCl at rates of 0, 25, 50, 100, 150 and 200 and 250 g K tree[-1] the 1st year and supplemented with 120 g N tree[-1] as ammonium sulfate (20% N) and 52 g P tree[-1] as triple superphosphate (19% P). The N, P, and K rates were doubled in the 2nd year. The K rates were tripled the 3rd year and supplemented with 360 g N tree[-1] and 105 g P tree[-1]. The acidifying ammonium fertilizers may increase exchangeable acidity in both trials. The fertilizers were broadcast around the tree 40 cm from crown projection. Each plot comprised four trees each covering an area of 7 m x 5 m, for a total of 286 trees ha[-1]. The experimental setup was a randomized block design with four replications. Fresh fruit yields were measured 1-3 times wk[-1] from January to June, starting approximately 97 d after fruit set.

Soils were sampled annually after harvest at four locations per tree in the 0-20 cm and 20-40 cm layers where most of the root system is located, then composited per plot. Soil samples were air dried and analyzed for K, Ca, Mg and (H + Al) [61]. The K, Ca and Mg were extracted using exchange resins, quantified by atomic absorption spectrophotometry and reported as mmol$_c$ dm[-3]. Exchangeable acidity (H+Al) was quantified by the SMP pH buffer method [62] and computed using the equation of [63] to convert buffer pH into mmol$_c$ (H+Al) dm[-3] as follows:

$$(H + Al) = 10\exp{(7.76 + 1.053pH_{SMP})}, \quad R^2 = 0.98 \tag{8}$$

Cation exchange capacity (CEC) was computed as the sum of cationic species. Assuming a soil bulk density of 1 kg dm[-3], CEC averaged 5.4 cmol$_c$ kg[-1].

3.3. Results

3.3.1. Influence of the K fertilization on cationic balances in soil

As shown by scatter and ternary diagrams (Figure 8), the large ellipses, that represent the distribution of cationic balances in the 0-20 and 20-40 cm layers, overlapped. However, the small ellipses (Figure 8) representing the confidence region about means differed significantly. The [K | Ca, Mg, H+Al] balance was higher in the 0-20 cm layer, indicating that more K accumulated in the surface layer as a result of surface K fertilizer applications.

Soil test K and cationic balances were averaged between the beginning and the end of the growing season to represent average soil conditions. The soil indices were related to fresh fruit yield (Figures 9a and 9b). In Figure 9, data are means of 4 replicates and bars are least significant differences.

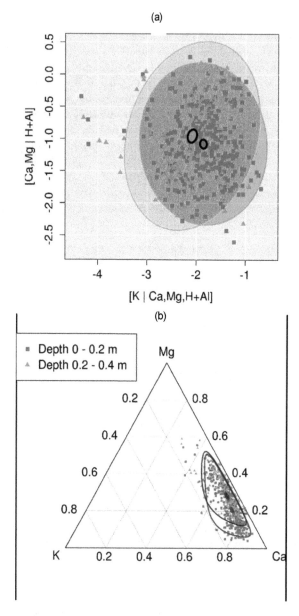

Figure 8. The K accumulated relatively more on the 0-20 cm layer following three years of K fertilization as shown by (a) two orthonormal cationic balances and (b) a K-Ca-Mg ternary diagram.

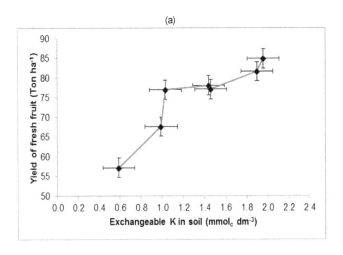

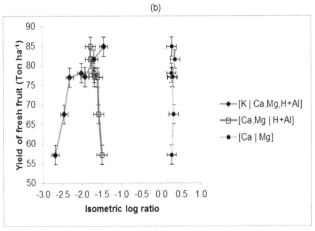

Figure 9. Exchangeable K (a), the [K | Ca,Mg,H+Al] balance (b) and fruit yield increased with added K. Other cationic balances did not change markedly.

3.3.2. Critical soil K concentration and balance in the N and K trials

The Cate-Nelson partitioning of the relationship between guava fresh fruit yield and either soil K level or the [K | Ca, Mg, (H+Al)] balance across the combined N and K fertilizer experiments indicates that the K level index classified two specimens as TN compared to four for the K balance index (Figure 10). The graphical representation of this soil-plant relationship indicates diagnostic advantage to using the K nutrient balance in rather than the K concentration.

The sensitivity, specificity, PPV and NPV criteria are presented in Table 3. We expect perform-ance criteria to be at least 80%. Low specificity indicates that some interactions with K leading to high yield, possibly involving Ca and Mg, have been ignored. Apparently, the *ceteris paribus* assumption did not apply to this study. The fact that the balance allows to adjust the K to other cationic species may account for failure to meet the *ceteris paribus* assumption.

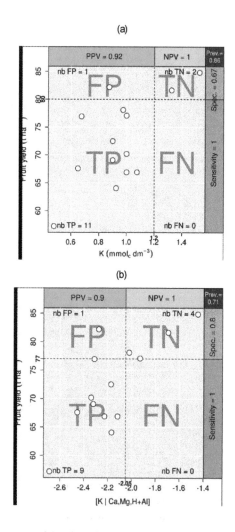

Figure 10. Cate-Nelson partitioning of the relationship between guava fresh fruit yield. critical values were (a) 1.2 mmol$_c$ K dm^{-3}. TN = 2; FN = 0; TP = 11; FP = 1 and (b) -2.07. TN = 4; FN = 0; TP = 9; FP = 1.

Soil K index	Sensitivity = TP/(TP+FN)	Specificity = TN/(TN+FP)	Positive predictive value = PPV=TP/(TP+FP)	Negative predictive value = NPV=TN/(TN+FN)
		%		
K level	100.0	66.7	91.7	100.0
K balance	100.0	80.0	90.0	100.0

Table 3. Performance of K indices in terms of sensitivity, specificity, PPV and NPV

4. Multi-element Balances in plant nutrition

4.1. Sequential binary partition

Plant nutrients are classified as essential macronutrients measured in % (N, S, P, Mg, Ca, K, Cl), essential micronutrients measured in mg kg^{-1} (Mn, Cu, Zn, Mo, B) and beneficial nutrients generally measured in mg or µg kg^{-1} but occasionally in % (Si, Na, Co, Ni, Se, Al, I, V) [64, 65, 15]. The plant ionome is defined as elemental tissue composition as related to the genome [66]. A subcomposition of plant ionome could be defined by the following simplex for conducting statistical analysis:

$$S^D = \mathcal{C}(C, N, P, K, Ca, Mg, B, S, Cl, Cu, Zn, Mn, Fe, Mo, F_v) \tag{9}$$

Where F_v is the filling value between 1000 g kg^{-1} and the sum of analytical data and $D = 15$, the total number of components including F_v. An SBP scheme can be elaborated based on well documented roles and stoichiometric rules provided by [17, 14, 12], who reported a large number of dual and multiple nutrient interactions in plants such as:

• Macronutrients have a stoichiometric relationship with carbon uptake;

• N with S, P, K, Ca, Mg, Fe, Mn, Zn, and Cu;

• NH4 with K, Ca, and Mg;

• S with N, P, Fe, Mn, Mo;

• P with N, K, Ca, Mg, B, Mo, Cu, Fe, Mn, Al, and Zn;

• Cl with N and S;

• K with N, P, Ca, Mg, Na, B, Mn, Mo, and Zn;

• Ca with N, K, Mg, Na, Cu, Fe, Mn, Ni, and Zn;

- Mg with N, P, B, Fe, Mn, Mo, Na, and Si;

- B with N, P, K, and Ca;

- Cu with N, P, K, Ca, Fe, Mn, and Zn;

- Fe with N, P, Ca, Mg, Cu, Mn, Co, and Zn;

- Zn with N, P, K, Ca, Mg, S, Na, Zn, Fe, and Mn;

- Mn with N, P, K, Ca, Mg, B, Mo, Ni, and Zn;

- Mo with N, P, K, S, Fe, and Mn.

4.2. Datasets

The tissue composition can be altered by environmental and seasonal factors. A dataset of 1909 potato (*Solanum tuberosum* L. cv. 'Superior') yields and ionomes was collected at five developmental stages between 1987 and 2002 in Quebec, Canada. The first mature leaf from top was sampled at 20-cm height (n = 502), bud stage (n = 544), beginning of flowering (n = 587), full bloom (n = 213) and fast tuber growth (n = 63) and analyzed for N, P, K, Ca, and Mg. The plant nutrient signatures at each developmental stage were compared using box-plots and discriminant analysis.

A critical hyper-ellipsoid can be viewed as a particular zone of the nutrient balance space where the probability to obtain high yield is high enough to satisfy the practitioner. The points lying inside the hyper- ellipsoid would be qualified as "balanced", and those lying outside the multi-dimensional construct, as "imbalanced". The practitioner might delineate intermediate zones if needed. Fertilizer trials were conducted to monitor balance change toward optimum nutrient conditions defined by the critical ellipses. In a P trial, P treatments applied to a P deficient soil were 0, 33, 66, 98 and 131 kg P ha^{-1}. In a K trial, K treatments of 0, 50, 100 and 150 kg K ha^{-1} were applied to a K deficient soil. The diagnostic leaf of potato was sampled at the beginning of flowering [67].

4.3. Seasonal change in nutrient compositions

The boxplots and the CoDa dendrogram illustrate the center and dispersion of nutrient balances per development stage (Figure 11). The [N, P, K | Ca, Mg] balance tended to decrease markedly during the season while the [N | P] and [Ca | Mg] balances tended to increase, and the [N,P | K] balance tended to decrease. The fast decrease in [N,P,K | Ca, Mg] balance is attributable to more N, P and K than Ca and Mg being transferred toward growing leaves during exponential growth and toward tubers during maturation. The K was more affected than N and P.

The discriminant scores (dots) and eigenvectors, as well as confidence regions at 95% level delineated the distributions of populations (large grey ellipses) and means (small color colored ellipses) across stages of plant development (Figure 12). The first axis, dominated by the Redfield [N | P] balance followed by the [N, P, K | Ca, Mg] balance, captured 92% of total inertia. It is noteworthy that the nutrient balance changed orderly from one developmental to the other. The *ilrs* can thus be described by trend equations and sample composition be detrended toward a specific developmental stage for diagnostic purposes. The seasonally increasing N/P ratio may indicate possible N or P imbalance at some point in time assuming a stationary N:P stoichiometric rule. However, the N/P ratio was found to vary widely between plant species during plant development, depending on relative growth rates [38]. The Redfield N/P ratio in eukaryotic microbes is a balance between two fundamental processes, protein and rRNA synthesis, resulting in a stable biochemical attractor toward a given protein: to RNA ratio [68]. The N/P ratio of plant biomass is used as indicator of N or P limitation but critical N/P ratios change with age and function of tissues [38]. Immature leaves of young plants assimilate and grow simultaneously and their demand for N and P follows the stoichiometryic rules of basic biochemical processes such as photosynthesis, respiration, protein synthesis, DNA duplication and transcription; growth becomes restricted to active meristems such as young leaves, shoot tips and inflorescences when plants get older [38]. Mature leaves are still photosynthetically active but no longer grow, which greatly reduces the P requirements for RNA and increases the N/P ratio. Nucleic-acid P can be mobilized from older leaves and transferred to younger leaves, leading to higher N/P ratios in older leaves [69], such as the first mature leaves of potatoes used as diagnostic tissue [67].

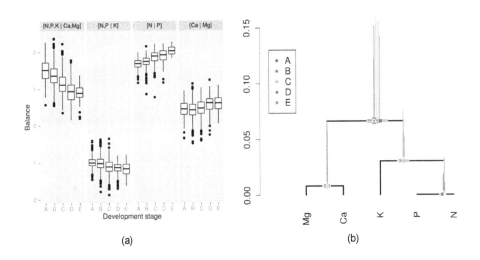

Figure 11. The (a) boxplots and (b) coda dendrogram of the four balances for five development stages.

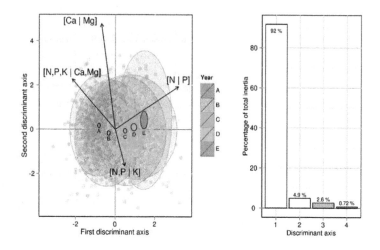

Figure 12. Discriminant analysis shows that nutrient balance changes orderly between the early and late developmental stages.

4.4. Defining reference balances for diagnostic purposes

The confidence region of optimum nutrition was defined by a 4-dimensional hyper-ellipsoid (Figure 13).

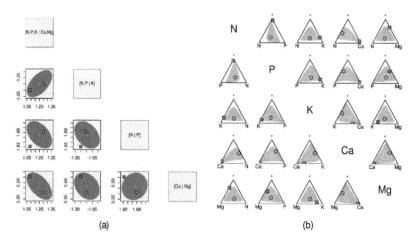

(a)   (b)

Figure 13. The ellipses define the optimum conditions of potato nutrition as found in surveyed high-yield potato stands (95% confidence regions about mean) illustrated as (a) balances and (b) scaled and centered concentrations.

The green and red points in Figure 13 represent specimens showing balanced and imbalanced nutrition, respectively. The fertilization of the potato should move nutrient signature toward the hyper-ellipsoid center. Added P perturbed the internal nutrient balance of cv. 'Superior' growing on a P deficient soil (Figure 14). The P trial showed that an addition of 98 kg P ha^{-1} allowed the balance to penetrate into the critical ellipse.

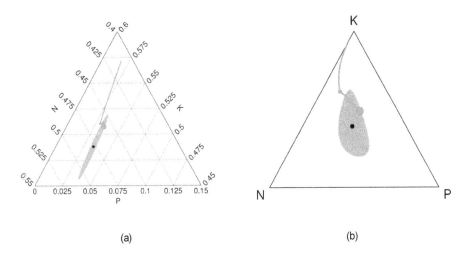

(a)   (b)

Figure 14. The P treatments applied to a P-deficient soil at rates of 0, 33, 66, 98 and 131 kg P ha^{-1} (rates increasing as red circles enlarged) perturbed the potato ionome that moved toward the critical ellipse. The ternary diagram scaled on proportions on the left (A) was centered and rescaled on the right (B) for better appreciation of the shape of the ellipse and the trend of the fertilization effect.

In Figure 15, it can be observed that added K also perturbed the nutrient balance: the potato ionome moved toward the critical ellipse. The 2nd K rate moved the K deficient plant ionome closer to the critical ellipse, but Ca shortage maintained the crop outside the critical ellipse. From the second application rate up, the perturbation was small. In this case, the Ca was likely to be the most limiting nutrient as shown on the ternary diagram.

The perturbation on 5 nutrients can be illustrated by a matrix of ternary diagrams (Figure 16). These diagrams show 2 nutrients and an asterisk (*) representing the sum of the 3 other components. The central dot is the mean of high yielders surrounded by its 95% confidence region represented by a black line.

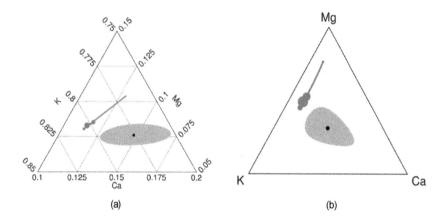

(a)   (b)

Figure 15. Ternary diagrams showing K treatments applied to a K-deficient soil at rates of 0, 50, 100 and 150 kg K ha^{-1} (rates increasing as red circles enlarged (a) zoomed on proportions and (b) centered and scaled.

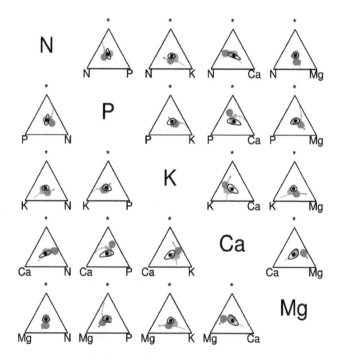

Figure 16. General view of plant ionomes in P and K deficient soils moving toward the reference ellpse for highly productive agroecosystems as P (blue) and K (red) fertilizers are added.

5. Compositional modeling of C mineralization of organic residues in soils

The carbon, nitrogen, phosphorus and sulfur cycles are interconnected in agroecosystems and often expressed using stoichiometric rules [4]. The ratio between total C and total N is the most simplified rule used in C mineralization studies but the $C_{organic}/N_{organic}$ and lignin/$N_{organic}$ ratios are also common. However, several biochemical components of organic matter are omitted in most studies, resulting in loss of information on the system. There are few studies on the relationship between labile or recalcitrant C and the biochemical composition of organic residues added to soil. [70] analyzed ash and N contents as well as four C fractions in organic residues representing pools of increasing resistance to decomposition. In this section, we related labile C in organic residues to this 6-part compositional vector of organic residues. The components were expressed as fractions on dry weight basis to compute a biological stability index using multiple linear regression models. The compositional vector was defined as follows:

$$S^6 = \mathcal{C}(SOL\,,\,HEM,\,CEL\,,\,LIG,\,N,\,Ash) \tag{10}$$

Where SOL = soluble matter, HEM = hemicellulose, CEL = cellulose, L IG= lignin, and N = total nitrogen.

Because scale dependency induces spurious correlations [71, 72, 73] and linear regression models are solved based on correlations between variables, the interpretation of regression coefficients is scale-dependent. To illustrate the problem of spurious correlations, chemical fractions were scaled on organic mass basis and analyzed using multiple linear regression.

The balance scheme reflected the C/N ratio and the order of decomposability of biochemical components (Figure 17). The SOL fraction was isolated from other biochemically labile fractions because its composition is complex, possibly including sugars, amino-sugars, amino-acids, and polypeptides as well as more recalcitrant or bacteriostatic easily solubilized polyphenols such as fulvic acids, tannic substances, resins, intermediate products, etc. The balance scheme was formalized by SBP as shown in Table 4.

Ilr balance	SOL	HEM	CEL	LIG	Total N	Ash	r	s
[SOL,HEM,CEL,LIG,N	Ash]	1	1	1	1	1	-1	5	1
[SOL,HEM,CEL,LIG	N]	1	1	1	1	-1	0	4	1
[SOL,HEM,CEL	LIG]	1	1	1	-1	0	0	3	1
[SOL	HEM,CEL]	1	-1	-1	0	0	0	1	2
[HEM	CEL]	0	1	-1	0	0	0	1	1

Table 4. Sequential binary partition of the biochemical composition of organic residues in Thuriès et al. (2002)

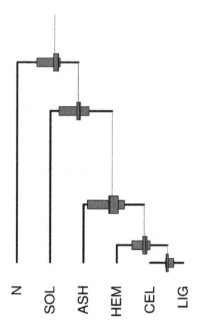

Figure 17. Graphical representation of the balances for the sequential binary partition defined in Table 3. SOL = soluble matter, HEM = hemicellulose, CEL = cellulose, LIG = lignin, N = total nitrogen, and ash.

The linear regression models relating labile C to bio-chemical fractions or balances showed R^2 values between 0.86 and 0.92 (Figure 18). For the 6-part (dry mass basis) and 5-part (organic matter basis) models, variation in labile C mesaured as evolved CO_2 was explained in part by total N and SOL as follows:

$$C_{labile} = 0.0773 + 0.5202SOL - 0.2515HEM - 0.3372CEL - 0.2882LIG + 2.3884N_{total} \qquad (11)$$

$$C_{labile} = -0.1776 + 0.5585SOL + 0.1909HEM + 2.0187N_{total} \qquad (12)$$

However, Equations 11 and 12 were subcompositionally incoherent. The intercept and the β coefficient for HEM showed opposite signs in equations 11 and 12 while CEL and LIG were absent in Equation12. This incoherence is attributable to spurious correlations (Table 5). Pearson correlation coefficients among raw proportions were not consistent in terms of value, significance or sign whether the proportions were expressed on the dry mass of the organic product (including ash) or on organic matter (LOI) basis.

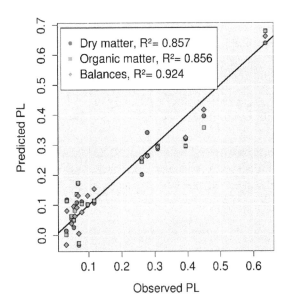

Figure 18. Regression analysis of C mineralization on dry matter (six raw components) or organic matter (five components) basis and using balances that reflects the order of decomposability of b

Component	SOL	HEM	CEL	LIG	Ash
			Pearson correlation coefficient		
			Dry matter basis (including ash)		
Total N	0.241	0.354	-0.462	-0.320	-0.184
SOL		-0.115	-0.232	-0.669	-0.027
HEM			-0.292	-0.293	-0.340
CEL				0.465	-0.495
			Organic matter basis (loss on ignition)		
Total N	0.466	0.067	-0.637	-0.475	-
SOL		-0.194	-0.425	-0.756	-
HEM			-0.409	-0.383	-
CEL				0.376	-

SOL = soluble substances; HEM = hemicelluloses; CEL = cellulose; LIG = lignin and cutin

Table 5. Scale dependency of the correlations between biochemical components in Thuriès et al. (2002); for 17 observations, $r = 0.468$ at $P = 0.05$ and $r = 0.590$ at $P = 0.01$

On the other hand, the labile C pool was largely explained by the *ilr* balances between C sources and total N, a surrogate of the C/N ratio, the balance between labile and refractory C sources, and between two labile C pools, one being more labile (HEM) than the other (CEL). The equation was as follows:

$$C_{labile} = 0.3065 - 0.1251[SOL\ .HEM,\ CEL\ ,\ LIG\ |\ N] + 0.0301[[SOL\ ,\ HEM,\ CEL\ |\ LIG]$$
$$+ 0.0019[SOL\ |\ HEM,\ CEL\ ] - 0.1063[HEM\ |CEL\ ] \tag{13}$$

Equation 13 shows that labile C increases with total N and higher proportions of more labile over more recalcitrant C forms. These findings indicate that the *ilr* coordinates provide a coherent interpretation of the C dynamics of organic products. The *ilrs* are not redundant, scale-invariant and free from spurious correlations.

6. Conclusions

This paper shows that the specific numerical properties of compositional data require log ratio transformations before conducting statistical analyses of soil and plant compositional data. Compared to raw concentration data, the orthonormal balances can be interpreted consistently and without numerical bias as isometric log ratio coordinates. The *ilr* approach can provide unbiased indices of nutrient balance in soils and plant tissues, biological stability of organic residues and soil quality. Well supported by techniques developed by compositional data analysts, the balance paradigm and the elaboration of its SBP schemes prompt that many concepts inherited from the past centuries be debated and revisited in soil fertility and plant nutrition.

Acknowledgements

The balance paradigm was elaborated within the plant nutrition and soil carbon modules of the research project entitled 'Implementing means to increase potato ecosystem services' (CRDPJ 385199 – 09). We acknowledge the financial support of the Natural Sciences and Engineering Research Council of Canada (NSERC), the Fundação de Amparo à Pesquisa do Estado de São Paulo (FAPESP), the Coordinação de Aperfeiçoamento de Pessoal de Nivel Superior (CAPES), as well as farm partners as follows: Cultures Dolbec Inc., St-Ubalde, Québec, Canada; Groupe Gosselin FG Inc., Pont Rouge, Québec, Canada; Agriparmentier Inc. and Prochamps Inc., Notre-Dame-du-Bon-Conseil, Québec, Canada; Ferme Daniel Bolduc et Fils Inc., Péribonka, Québec, Canada.

Author details

S.É. Parent[1], L.E. Parent[1], D.E. Rozanne[2], A. Hernandes[3] and W. Natale[3]

*Address all correspondence to: Leon-Etienne.Parent@fsaa.ulaval.ca

1 Department of Soils and Agrifood Engineering, Université Laval, Québec (Qc), Canada

2 Departamento de Agronomia, Unesp, Universidade Estadual Paulista, Campus de Registro, Registro, São Paulo, Brasil

3 Departamento de Solos e Adubos, Unesp, Universidade Estadual Paulista, Jaboticabal, São Paulo, Brasil

References

[1] Sims, T. Soil fertility evaluation. In Sumner ME (ed.). Handbook of soil science. Boca Raton FL: CRC Press, 2000. p. D-113-D-153.

[2] Epstein W., Bloom AJ. Mineral nutrition of plants: Principles and Perspectives. Sunderland MA: Sinauer Ass. Inc.; 2005.

[3] Gliessman SR. Agroecology. Ecological processes in sustainable agriculture. Chelsea, MI: Ann Arbor Press; 1999.

[4] Stevenson F. Cycles of soil. Carbon, nitrogen, phosphorus, sulfur, micronutrients. NY: Wiley Intersci.; 1986.

[5] Lahner B, Gong J, Mahmoudian M, Smith EL, Abid KB, Rogers EE, Guerinot ML, Harper JF, Ward JM, McIntyre L, Schroeder, JI, Salt DE. Genomic scale profiling of nutrient and trace elements in *Arabidopsis thaliana*. Nature Biotechnology 2003;21: 1215–1221.

[6] Williams L, Salt DE. The plant ionome coming into focus. Current Opinion in Plant Biology 2003;12(3) 247–249.

[7] Han WX, Fang J Y, Reich PB, Woodward FI, Wang ZH. Biogeography and variability of eleven mineral elements in plant leaves across gradients of climate, soil and plant functional type in China. Ecology Letters 2011;14: 788–796.

[8] Schlotter D, Schack-Kirchner H., Hildebrand EE, von Wilpert K. Equivalence or complementarity of soil-solution extraction methods. Journal of Plant Nutrition and Soil Science 2012;175(2) 236-244.

[9] Gabrielle B, Da-Silveira J, Houot S, Francou C. Simulating urban waste compost effects on carbon and nitrogen dynamics using a biochemical index. Journal of Environmental Quality 2004;33: 2333-2342.

[10] Lashermes G, Nicolardot B, Parnadeau V, Thuriès L, Chaussod R, Guillotin ML, Li-nières M, Mary B, Metzger L, Morvan T, Tricaud A, Villette C, Houot S. Indicators of potential residual carbon in soils after exogenous organic matter application. European Journal of Soil Science 2009;60: 297-310.

[11] Paul EA, Clark FE. Soil biology and biochemistry. NY: Academic Press; 1989.

[12] McLean EO. Contrasting concepts in soil test interpretation: sufficiency levels of available nutrients versus basic cation saturation ratios. In M, Stelly (ed.) Soil testing: Correlating and interpreting the analytical results. Madison WI: American Society of Agronomy Special Publication 29; 1984. p. 39-54.

[13] Walworth JL, Sumner ME. The Diagnosis and Recommendation Integrated System (DRIS). Advances in Soil Science 1987;6: 149-188.

[14] Bergmann W. Ernährungsstörungen bei Kulturpflanzen. 2. Auflage. Stuttgart: Gustav Fisher Verlag; 1988.

[15] Marshner H. Mineral nutrition of higher plants. NY: Academic Press; 1995.

[16] Malavolta E. Manual de nutrição de plantas. Ceres, Piracicaba; 2006.

[17] Ingestad T. New concepts on soil fertility and plant nutrition as illustrated by research on forest trees and stands. Geoderma 1987;40: 237-252.

[18] Körner C. The grand challenges in functional plant ecology. Frontiers in Plant Science 2011;2: 1.

[19] Geraldson CM. Intensity and balance concept as an approach to optimal vegetable production. Communications in Soil Science and Plant Analysis 1970;1(4) 187-196.

[20] Geraldson CM. (1984). Nutrient intensity and balance. in Stelly, M. (ed) Soil testing: Correlating and Interpreting Analytical Results. Madison WI: American Society of Agronomy Special Publication 29; 1984.,p. 75-84.

[21] Liebhardt WC. The Basic Cation Saturation Ratio Concept and Lime and Potassium Recommendations on Delaware's Coastal Plain Soils. Soil Science Society of America Journal 1981;52:1259-1264.

[22] Kopittke PM, Menzies NW. A Review of the Use of the Basic Cation Saturation Ratio and the "Ideal" Soil Soil Science Society of America Journal 2007;71: 259-265.

[23] Lagatu H, Maume L. Le diagnostic foliaire de la pomme de terre. Annales de l'École Nationale Agronomique de Montpellier (France) 1934;22: 50-158.

[24] Egozcue JJ, Pawlowsky-Glahn V. Groups of parts and their balances in compositional data analysis. Mathematical Geosciences 2005;37: 795-828.

[25] Beaufils ER. Diagnosis and recommendation integrated system (DRIS). Soil Science Bulletin, 1, Pietermaritzburg: University of Natal; 1973.

[26] Aitchison J. The Statistical Analysis of Compositional Data. London: Chapman and Hall; 1986.

[27] Egozcue JJ, Pawlowsky-Glahn V, Mateu-Figueras G, Barceló-Vidal C. Isometric log-ratio transformations for compositional data analysis. Mathematical Geology 2003;35: 279-300.

[28] Roy RN, Misra RV, Lesschen JP, Smaling AN.. Assessment of soil nutrient balance. Approaches and Methodologies. Rome: FAO fertilizer and plant nutrition bulletin 14; 2003. http://www.fao.org/docrep/006/y5066e/y5066e00.htm (accessed 20 August 2012).

[29] Filzmoser P, Hron K. Robust statistical analysis. In Pawlowsky-Glahn V, Buccianti A (eds.) Compositional data analysis: Theory and Applications. NY: John Wiley and Sons; 2011. p. 57-72.

[30]] Pawlowsky-Glahn V., Mateu-Fugueras, G., and Buccianti A.. Compositional data analysis: Theory and Applications. London: The Geological Society of London; 2006.

[31] Pawlowsky-Glahn V, Buccianti A. Compositional data analysis: Theory and Applications. NY: John Wiley and Sons; 2011.

[32] Filzmoser P, Hron K, Reimann C. Univariate statistical analysis of environmental (compositional) data: Problems and possibilities. Science of Total Environment 2009;407(23) 6100-8.

[33]] Tolosana-Delgado R, van den Boogart, KG. Linear models with compositions in R. In Pawlowsky-Glahn V, Buccianti A (eds.) Compositional data analysis: Theory and Applications. NY: John Wiley and Sons; 2011. p. 356-371.

[34] Parent, LE, Dafir M. A theoretical concept of compositional nutrient diagnosis. Journal of the American Society for Horticultural Science 1992;117: 239-242.

[35] Parent, LE, Cissé ES, Tremblay N, Bélair G. Row-centred logratios as nutrient indexes for saturated extracts of organic soils. Canadian Journal of Soil Science 1997;77: 571-578.

[36] Lopez J, Parent LE, Tremblay N, Gosselin A. Sulfate accumulation and Ca balance in hydroponic tomato culture. Journal of Plant Nutrition 2002;25(7) 1585-1597.

[37] Pawlowsky-Glahn V, Egozcue JJ. Exploring Compositional Data with the CoDa-Dendrogram. Austrian Journal of Statistics 2011;40(1&2) 103-113.

[38] Güsewell S. N:P ratios in terrestrial plants: variation and functional significance. New Phytologist 2004;164: 243-266.

[39] Egozcue JJ., Daunis-i Estadella J, Pawlowsky-Glahn V, Hron K, Filzmoser P. Simplicial regression. The normal model. Journal of Applied Probability and Statistics 2012;6(1&2) 87-108.

[40] Buccianti A.Natural laws governing the distribution of the elements. In Pawlowsky-Glahn V, Buccianti A (eds.) Compositional data analysis: Theory and Applications. NY: John Wiley and Sons; 2011. p. 255-266.

[41] Parent, LE. Diagnosis of the nutrient compositional space of fruit crops. Revista Brasileira Fruticultura 2011;33: 321-334.

[42] Parent LE, Parent S.É, Rozane DE, Amorim DA, Hernandes A, Natale W. Unbiased approach to diagnose the nutrient status of guava. Acta Horticulturae 2012 (in press).

[43] Abdi, D, Ziadi N, Parent, LE. Compositional analysis of phosphorus pools in Canadian Mollisols. Proceedings of the 4th International Workshop Compositional Analysis, 9-13 May 2011, San Feliu-de-Guixols, Girona, Spain. http://www.ma3.upc.edu/users/ortego/codawork11-Proceedings/Admin/Files/FilePaper/p57.pdf (accessed 20 August 2012).

[44] Parent LE, Parent, S-É, Kätterer T, Egozcue JJ. Fractal and compositional analysis of soil aggregation., Girona, Spain: Proceedings of the 4th International Workshop Compositional Analysis, 9-13 May 2011, San Feliu-de-Guixols, Girona, Spain. http://www.ma3.upc.edu/users/ortego/codawork11-Proceedings/Admin/Files/FilePaper/p5.pdf (accessed 20 August 2012).

[45] Parent LE, de Almeida C X, Hernandes A, Egozcue JJ, Gülser C, Bolinder MA, Kätterer T, Andrén O, Parent S.-É, Anctil F, Centurion JF, Natale W. Compositional analysis for an unbiased measure of soil aggregation. Geoderma 2012;179–180: 123-131.

[46] Rodgers, J. L., Nicewander, W. A. and Toothaker, L. 1984. Linearly independent, orthogonal, and uncorrelated variables. Amer. Statist. 38,133-134.

[47] Egozcue JJ, Pawlowsky-Glahn, V. Simplicial geometry for compositional data. In Compositional Data Analysis in the Geosciences: From Theory to Practice. Geological Society (London, England), 145-159.

[48] Lovell D, Müller W, Tayler J, Zwart A, Helliwell C. Proportions, percentages, ppm: do the molecular biosciences treat compoitional data right? In Pawlowsky-Glahn V, Buccianti A (eds.) Compositional data analysis: Theory and Applications. NY: John Wiley and Sons; 2011. p. 193-207.

[49] Parent LE, Natale W, Ziadi, N. Compositional Nutrient Diagnosis of Corn using the Mahalanobis Distance as Nutrient Imbalance Index. Canadian Journal of Soil Science 2009;89: 383-390.

[50] Hadi A. Identifying multiple outliers in multivariate data. Journal of the Royal Statistical Society1992;B54: 761_771.

[51] Hadi A. A modification of a method for the detection of outliers in multivariate samples. Journal of the Royal Statistical Society1994;B56: 393-396.

[52] Nelson LA, Anderson RL. Partitioning of soil test-crop response probability. in M. Stelly (ed.) Soil testing: Correlating and interpreting the analytical results. Madison WI; American Society of Agronomy Special Publication 29I; 1984. P. 19-38.

[53] MedStats Club. Statistics tutorials, calculators and tools (2012). http://www.medstats.org/sens-spec Accessed 20 August 2012.

[54] Lander ES, Botstein, D. Mapping Mendelian Factors Underlying Quantitative Traits Using RFLP Linkage Maps. Genetics 1989;121: 185-199.

[55] R Development Core Team. R: A Language and Environment for Statistical Computing. R Foundation for Statistical Computing Version 2.13.1; 2011. http://www.R-project.org (accessed 20 August 2012).

[56] van den Boogaart KG, Tolosana-Delgado R,, Bren, M. compositions: Compositional Data Analysis. R package version 1.10-2 ; 2012. http://cran.r-project.org/web/packages/compositions/compositions.pdf (accessed 20 August 2012).

[57] Thode HC Jr. Testing for Normality. NY: Marcel Dekker; 2002.

[58] Gross J. nortest: Five omnibus tests for the composite hypothesis of normality. R package version 1.0 ; 2006. http://CRAN.R-project.org/package=nortest (accessed 20 August 2012).

[59] Filzmoser, P, Gschwandtner, M. mvoutlier: Multivariate outlier detection based on robust methods. R package version 1.9.4 ; 2011. http://CRAN.R-project.org/package=mvoutlier (accessed 20 August 2012).

[60] Embrapa. Sistema Brasileiro de classificação de solos. 2nd Ed., Rio de Janeiro: Embrapa Solos; 2006.

[61] Raij B van, Quaggio JA, Cantarella H, Ferreira ME, Lopes AS, Bataglia OC. Análise química do solo para fins de fertilidade. Campinas: Fundação Cargill; 1987.

[62] Shoemaker HE, Mclean EO, Pratt PF. Buffer methods for determining lime requirement of soils with appreciable amounts of extractable aluminium. Soil Science Society of Amica Proceedings 1961;25: 274-277.

[63] Quaggio JA, van Raij B., Malavolta E. Alternative use of the SMP-buffer solution to determine lime requirement of soils. Communications in Soil Science and Plant Analysis 1985;16: 245-260.

[64] Callot G, Chamayou H, Maertens C, Salsac L. Mieux comprendre les interactions sol-racine: incidence sur la nutrition minérale. Paris : INRA; 1982.

[65] Smith GS, Buwalda JB., Clark CJ. Nutrient dynamics in a kiwifruit ecosystem. Scientia Horticulturae 1988;37: 87-109.

[66] Salt ED, Baxter I, Lahner B. Ionomics and the study of the plant ionome. Annual Review of Plant Biology 2008;59: 709–33.

[67] Parent LE, Cambouris AN, Muhawenimana, A. Multivariate diagnosis of nutrient imbalance in potato crops. Soil Science Society of America Journal1994;58:1432-1438.

[68] Loladze I., Elser, JJ. The origins of the Redfield nitrogen-to-phosphorus ratio are in a homoeostatic protein-to-rRNA ratio. Ecology Letters 2011;14: 244-250.

[69] Usuda H. Phosphate deficiency in maize. V. Mobilization of nitrogen and phosphorus within shoots of young plants and its relationship to senescence. Plant Cell Physiology 1995;36: 1041–1049.

[70] Thuriès L, Pansu M, Larré-Larrouy MC., Feller C. Biochemical composition and mineralization kinetics of organic inputs in a sandy soil. Soil Biology and Biochemistry 2002;34: 239-250.

[71] Pearson K. Mathematical contributions to the theory of evolution. On a form of spurious correlation which may arise when indices are used in the measurement of organs. Proceedings of the Royal Society 1897;LX: 489-502.

[72] Tanner J. Fallacy of per-weight and per-surface area standards, and their relation to spurious correlation. Journal of Physiology 1949;2: 1-15.

[73] Chayes F. On correlation between variables of constant sum. Journal of Geophysical Research 1960;65: 4185-4193.

Chapter 6

Soil Acidity and Liming in Tropical Fruit Orchards

William Natale, Danilo Eduardo Rozane,
Serge-Étienne Parent and Léon Etienne Parent

Additional information is available at the end of the chapter

1. Introduction

In Brazil, agribusiness generates some US\$ 330 billion in revenue per year and is the most important sector of country's economy, accounting for 30% of GDP, 36% of exports and 37% of the jobs. This activity is one of the main reasons for Brazil's trade surplus in recent years, with annual farm exports worth more than US\$ 60 billion [1].

Based on these figures, it is no exaggeration to say that more than one-third of the Brazilian annual wealth is supported by a single natural resource: the soil. This justifies the importance of studies on how to preserve and improve this valuable resource. According to [2], soils have natural limits to their ability to nourish plants and sustain crop productivity. The degradation of soil quality reduces this ability and at the same time deteriorates the quality of water for various uses. It is senseless to claim that agricultural technology can compensate for poor soil management.

Fruit growing is an important component of Brazilian agriculture, occupying 2.3 million hectares and producing 41 million metric tons of fruit annually, totalling some US\$ 10 billions. This ranks Brazil among the world's leading fruit producers [3]. Despite this standout position, fruit yield remains unsatisfactory compared to many other countries. Among factors contributing to this situation, perhaps the most important is the deficient use of techniques to manage soils, crops and the environment.

Because of advances in genetic improvement in recent decades, plants now produce more yields of higher-quality fruits, but the demand for and the export of nutrients, as can be expected, are also higher. On the other hand, Brazilian soils tend to be naturally acic and low in fertility and/or are subjected to overexploitation, leading to exhaustion. Soil acidity is one of the main factors that reduce crop yields as in other tropical regions of the globe. Liming is a widely used technique in annual cropping systems but for perennials such as fruit trees, liming is more complicated due to the characteristics of these plants and the lack of scientific knowledge on this subject. Fruit trees, like all other perennials, keep producing for many years in practically the same volume of soil, which is the reason why soil acidity requires special attention. Despite the high importance of lime application for most fruit trees, there is a lack of information on the effects of this soil treatment technique during the planting, formation and production stages of orchards.

Therefore, it is important to study the effects on orchards of soil acidity correction, especially through liming, by monitoring soil chemistry and the response of the trees. Better knowledge in this respect can improve fruit crop productivity that translates into higher profits for farmers.

2. Soil acidity and liming

The soil, from where mankind has drawn its main sustenance since the beginning of the civilization, requires adequate management to maintain its fertility and nutrient availability sufficient to sustain the fundamental role of crops in supporting human life.

Among soil environmental factors, acidity (pH, base saturation, potential acidity and nutrient solubility) is the one that affects most crop yields, particularly in tropical regions [4]. According to [5], the low fertility found in acidic soils is strongly associated with deficient levels of exchangeable bases and excessive amounts of of aluminum and manganese. The application of fertilizers that acidify the soil aggravates this problem, unless a well-planned liming program is implemented.

Some soils are naturally acidic due to relative shortage of basic cations in the original material or to processes that causes the loss of elements like potassium, calcium and magnesium [6]. Other soils, although not originally acid, become so due to the removal of exchangeable cations from the surface of colloids, caused by: a) rainwater; b) alteration of clay minerals; c) ion exchange of roots; d) decomposition of organic matter; and e) addition of nitrogen fertilizers.

Although liming is recognized as a beneficial practice to reduce soil acidity, it is often not employed, or is conducted inadequately. Limestone raises soil pH, neutralizes toxic aluminum and supplies calcium and magnesium to the crops. These factors promote the development of root systems and enhance the use of nutrients and water by the plant [7]. In soils of tropical regions, acid reaction and low levels of basic cations such as calcium are ever-lasting problems. Under these conditions, the application of limestone is an inexpensive, fast and efficient way to tackle both problems [8].

Among Brazilian minerals, limestone occupies first place: the country has estimated reserves of some 53 billion metric tons well distributed throughout the country and generally of good quality, making it a relatively inexpensive agricultural input. However, despite the abundance of limestone and the need for liming, this soil corrective measure is not used at a sufficient extent in the majority of Brazilian farming regions [5].

The application of limestone on annual crops, with homogeneous incorporation in the soil, is a common practice, although not recognized as it should be. In perennial crops, the incorporation of limestone ismore complex due to the intrinsic characteristics of these plants and the lack of scientific and technological information [9]. This is the case, for example, for the majority of fruit crops in Brazil.

In acid soils with high aluminum saturation, liming promotes the neutralization of the toxic Al in the surface layers, hence enabling more intense proliferation of roots with positive effect on plant growth. However, it is important to consider the need to incorporate the limestone thoroughly into the soil at the time of planting perennial crops because surface application alone acts slowly on the deeper soil layers and a soil insufficiently corrected at the establishment of the orchard can impair crop productivity for a long time [10]. The homogeneous incorporation of limestone allows greater contact between the amendment and the sources of acidity, speeding up the corrective effects that support efficient use of water and nutrients by the plant in the amended layer.

The importance of the root system is obvious because there is a close dependency between root development and the aboveground portion of the plant. The greater or lesser success of applying limestone and fertilizers, in turn, depends on the nature of the root system and on the volume of the soil effectively exploited by the particular plant species. Correction of acidity is the most efficient way to eliminate chemical barriers to the full development of the roots, and consequently, of the plant.

Unlike other agricultural inputs such as fertilizers, herbicides and insecticides, limestone can be considered an investment, because its benefits last over more than one harvest. This is due to the low solubility of the common limestones and the variability of particle sizes in crushed limestone, giving them different capacities to neutralize acidity over time. Therefore, two factors should be considered: the rate at which the acidity is corrected and the duration of the effects of liming. Fine particles promote rapid acidity correction, but this effect declines more quickly due to their faster solubilization. Therefore, the most efficient liming involves application of material with varied grain sizes to promote fast initial acidity correction with sufficient residual effect as well. The Brazilian law (2006) states that the reactivity of liming materials after a period of three months following soil application is zero for large particles more than 2 mm in diameter, 20% for particles in the range of 0.84 and 2 mm in diameter, 60% for particles from 0.30 to 0.84 mm in diameter, and 100% for fine particle less than 0.30 mm.

Because of the residual effect of limestone, liming materials applied to the soil at the time of planting orchard seedlings can keep the soil within acceptable acidity range for a certain period of time. However, determining the duration and intensity of the residual effect of liming at the moment of planting fruit orchards has not been widely studied, both due to experimental constraints and the time necessary to obtain satisfactory results [11-13].

Based on the above aspects, the best approach for liming is to apply limestone with larger grain size at the time of planting fruit orchards, with homogeneous incorporation in the soil, to prolong the residual effect, followed by use of fine material on adult trees, limited to the

surface, because the incorporation of corrective materials into the soil may induce phytosanitary problems to the plants. Materials with finer particles can move more easily through
the soil profile, correcting the acidity only in the surface layers.

Considering the perennial nature and cultivation conditions of fruit trees, the path of
limestone particles in the soil can vary along with various factors, including physical ones,
through the channels left by the decomposition of roots [14]. According to [15] and [16],
another explanation for particle flow through the soil profile is the formation of ion pairs
(Ca^{2+} and Mg^{2+}) and organic acids (RO⁻ and RCOO⁻) of high solubility and low molecular
weight that can be leached to deeper layers. Besides these mechanisms, according to [17]
other compounds may form such as $Ca(HCO_3)_2$ and $Mg(HCO_3)_2$. Nitrogen fertilization, in
turn, can promote the formation of soluble salts, such as calcium nitrate, which percolate
down through the soil in forms dissolved in water [18]. According to [11], it is probable
that the sum of the contributions of all these processes is more important than each one
individually. Finally, the movement of these particles depends on the dose of the corrective measure employed, the time after application, soil type and the type of fertilization
applied to the orchard.

3. Sampling of leaves and soil in fruit orchards

The soils in Brazil are mainly tropical, are low in fertility and normally show acid reaction.
This is particularly due to their weathering during soil formation. This is one of the main
reasons for applying lime and fertilizers in farmed areas. Another important factor is that
because of Brazil's continental dimensions, it has a wide range of climate and soil characteristics, requiring different liming and fertilization regimes depending on orchard location.

Furthermore, different soils have different nutrient deficiencies, and plant species vary
greatly in their nutritional demands. Therefore, the only reliable way to identify the best
acidity corrective measure or fertilizer for use in a determined place for a particular crop is
soil analysis.

Besides soil analysis, which is a well-established practice in agriculture, it is also important
for the majority of fruit trees to analyze the leaves for determining the pattern of nutrient
uptake over time [19]. Because adult fruit trees require a certain degree of nutritional stability, leaf diagnosis allows adjusting fertilization programs over the years so as not to impair
fruit harvest in the same year. Hence, leaf sampling and soil analysis are useful methods to
monitor the effects of liming and fertilization on plant nutrition. Growing fruit trees is a
long-term activity where the plants continue exploiting practically the same volume of soil
for many years. In this situation, chemical impediments (acidity) as well as physical ones
(soil compaction) can reduce the efficiency of the roots in exploiting the soil. Therefore, the
only way to determine to what extent plants are using the nutrients applied via fertilization
and liming is to diagnose their nutritional status by leaf analysis.

Chemical analysis is the easiest and most practical way to assess soil fertility. Proper sampling is essential to obtain reliable results because if the samples are not representative, the

results do not accurately reflect the true soil fertility. Soil sampling is a common practice among farmers for annual crops, but it has not been widely studied for perennial crops such as fruit trees, raising doubts on its reliability. The present recommendations are to sample the area that receives soil treatment. However, some works have shown a higher correlation between leaf nutrient levels of fruit trees and soil nutrient levels in the paths between rows than in the rows [12, 20]. So, which soil samples should be analyzed, those from the treated area or between the rows? How should correlations be interpreted? And at what depth should the soil be sampled? These are difficult questions to answer, and according to [8], it also is not easy to design studies for this purpose.

At the time of planting fruit orchards, the soil sampling procedure is the same as for annual crops, namely across the entire representative area. In producing orchards, it is important to sample the region under the projection of the tree crowns, which is the area that usually receives fertilizers. Samples should be collected at the end of harvest from 20 points in each homogeneous plot (same cultivar, age, productivity, soil type, management and fertilization). At the same time, samples should be taken between the rows to measure lime requirements if necessary. Studies have shown that acidification occurs more intensely under the projection of tree crowns due to nitrogen fertilization, application of organic wastes and accumulation of plant material from pruning. As a rule, limestone is more often applied in strips under crown projection than onto areas between rows.

The most common method to calculate lime requirement in Brazil is base saturation [10]. The formula is as follows:

$$LN\ (ton\ h\,a^{-1}) = \frac{(V_2 - V_1) \times CEC}{TRNP \times 10}$$

where:

LN is the need for limestone, in ton ha^{-1};

V_2 is the target base saturation for the crop;

V_1 is base saturation of the soil;

CEC is soil's cation exchange capacity; and

TRNP is the total relative neutralization power, which considers the quantity of carbonates present in the limestone and lime's granulometry.

The soil layer typically sampled is the surface 0 to 20 cm. However, fruit trees exploit a much larger soil volume compared to annual plants, so it is important to analyze the properties of the deeper layers, especially regarding calcium and aluminum concentrations. This may lead to gypsum application, which neutralizes toxic Al and allows increasing Ca concentration in deeper layers, an important factor for the proliferation of the root system and its exploitation of a larger soil volume.

Gypsum is indicated, for crops in general, when analysis of the soil from the 20-40 cm layer reveals calcium concentrations lower than 4 mmol$_c$ dm^{-3} and/or aluminum saturation above 40%. The need for gypsum is estimated by the following equation [10]:

$GN\ (kg\ ha^{-1}) = 6 \times clay$

where:

GN is the need for gypsum, in kg ha^{-1}; and

Clay is the soil clay content, in g kg^{-1}.

For diagnosing the nutritional status of plants, leaf analysis is the most efficient technique, but also the one where errors occur most frequently. Each plant species has its own sampling procedure. Thus, the leaves to be collected and the time of the year are critical for successful diagnosis of the plant nutritional status. In the case of fruit trees, such as the guava tree, the third pair of leaves should be collected from each plant, with 25 pairs being collected from each homogeneous plot. This should be done at the time of full flowering [21]. These leaves should be immediately taken to the laboratory for washing, drying, grinding and analysis. The next step is to interpret the results, based on studies conducted under field conditions in high-yielding orchards.

In summary, when performed correctly, analysis of soil and leaf samples can allow making recommendations on liming and fertilization to improve fruit yield and quality, and hence increase the profits of fruit growers.

4. Considerations on fruit growing areas and fructiferous plants

By comparison with other regions of the world, Brazil is endowed with adequate characteristics of soil, climate, water availability and diversity of fruit species that give it great potential as a fruit producer and exporter.

These favorable conditions for the development of fruit growing regions expand agroindustrial activity and boost exports, not only because of the nutritive value of fruits, but also because of the perspective they represent to increase farm production. Besides this, cultivation of perennial species, as are most fructiferous plants, triggers the occupation of areas with soils classified unsuitable for the majority of annual crops.

Despite those advantages, there is still a lack of information on the correct management of soil fertility, the choice of inputs and the nutritional needs of fruit-bearing plants, preventing Brazil from realizing its full agribusiness potential.

Because fruit orchards are perrenials and trees' roots remain practically restricted to the same soil space for many years, it is important to incorporate homogeneous amounts of limestone to deep levels before planting so the root system can develop adequately for efficient uptake of water and nutrients. This enhances the development and nutritional status of the plants with less need for fertilizers, thus improving the cost-benefit ratio of inputs and boosting crop productivity. In producing orchards, repeatedly applied high doses of nitrogen over restricted areas (projection of the tree crowns) aggravate the problem of soil acidity, hence requiring regular soil chemical analysis.

Fruit growing is an important activity in most Brazilian regions, especially those covered with latosols (oxisols) and argisols (ultisols), which are generally deep and permeable, hence providing ideal conditions for perennial crops that produce a broad and well developed root system. Nevertheless, these soils have a strong acid reaction and are low in nutrients, so they need liming and fertilization.

Roots do not develop satisfactorily in highly acid soils. Among the acidity factors, aluminum toxicity and calcium deficiency are indicated as the most relevant restrictions to root growth. Regarding acidity correction with application of limestone at soil surface without no incorporation, research reported low movement of the lime to deeper layers, depending on the dose applied, the time elapsed and the fertilization regime.

Knowledge to support adequate management of soil fertility and plant nutrition is particularly important for fruit growing systems, given the influence these production factors on the qualities of fruits, such as color, size, taste, aroma and appearance. Meeting plants' nutrient needs is one of the key factors for the success of this activity, because besides affecting yield and quality, adequate nutrition increases plant growth, tolerance to pests and diseases and postharvest longevity. On the one hand, the demand of this group of plants for mineral elements is relatively high, while on the other tropical soils in which they are normally cultivated are low in nutrients, making it imperative to apply nearly all the nutrients necessary for their full development. This leads to the use of high amounts of fertilizers [21, 11] and corrective measures [12, 13, 22] in orchards and requires technical competence for economically rational use of these inputs. For Brazilian conditions, the analyses can be interpreted with the help of the Fert-Goiaba software [23].

Information on the effects of liming on the development and nutritional status of fructiferous plants is very limited in the literature despite the widespread scientific recognition of the importance of acidity correction. Much more attention has been paid to these aspects in annual cropping systems, but those findings cannot be transferred directly to perennial plants because the latter do not react the same way to liming and fertilization for a various reasons as follows [24]:

a. The roots of perennial plants like fructiferous ones exploit a large volume of soil, which increases as plant ages, and little is known about the nutritional reserves in deeper soil layers.

b. Roots, trunk, branches and leaves of perennial plants form large reservoirs of nutrients. Therefore, these plants have hidden early deficiency that makes it difficult for the farmer to implement corrective measures at the right time. Furthermore, it takes longer periods of time for these reservoirs to replenish, making these plants slower to react to nutrient applications.

c. The regular pruning of fructiferous plants complicates the problem of liming and fertilization. By restricting vegetative development, pruning hides even more deficiency symptoms and hence the effects of corrective measures. However, in many cases pruning is essential, because, as the saying goes, "hunger for light is just as harmful as hunger for nutrients."

d. Liming and fertilization of orchards are determinant not only in the current growing season, but also for harvests to come because the inputs applied at one time will also supply the pending production by promoting the formation of new fruit-bearing shoots and building up nutrient reserves in the roots and the aboveground biomass for the following seasons.

For a long time some fruit trees, especially those native to tropical regions like guava and carambola, were considered to be rustic plants, so their development was thought to be independent of edaphoclimatic conditions as is still felt today about pastures. However, it is not possible to imagine that a soil can be exploited by a crop indefinitely without replenishing the nutrient reserves or correcting the acidity. However, due to the characteristics of perennial fruit-bearing plants, the difficulties of conducting long-term experiments under present research funding support tend to discourage researchers.

5. Liming at planting and during formation of fruit orchards

Because soil acidity is one of the most important factors limiting farm production in tropical regions, an experiment was performed to assess the effects of liming on soil fertility as well as the mineral nutrition and yield of guava (*Psidium guajava*) [12]. The limestone (CaO=45,6% and MgO=10,2%) was incorporated into the soil in July and August 1999 and the orchard was planted four months later (December 1999) using the Paluma cultivar propagated by cuttings. The corrective measure was applied manually on the entire field area, half incorporated with a moldboard plow and the other half applied and incorporated later using a disk plow (both implements reaching a depth of 0–30 cm). The soil was a dystrophic red latosol with base saturation (V) of 26% in the 0–20 cm layer. The experimental design consisted of random blocks of five treatments and four repetitions. The limestone (reactivity = 94%) rates were 0, 1.85, 3.71, 5.56, and 7.41 ton ha^{-1}. Chemical analysis of the soil was carried out for 78 months after liming and nutrient status and tree productivity were evaluated during five consecutive harvests. Liming changed chemical attributes of the soil related to acidity down to a depth of 60 cm, raising the pH, Ca, Mg, sum of bases (SB) and base saturation (V) and diminishing potential acidity (H+Al).

During four years after orchard establishment, there was a significant correlation between leaf and soil Ca (Table 1). In general, the same pattern occurred for Mg, with higher correlation of leaf Mg levels with Mg concentration in the soil between the rows. This can indicate that with the exhaustion of these bases in the tree rows, the roots of guava trees could absorb nutrients effectively between the rows, indicating the importance of liming the entire area.

The yearly productivity during the experimental period (2002 to 2006) and the accumulated guava yields are presented in Figures 1a and 1b. Note the close fit of the production data to polynomial functions of the lime rates.

Soil nutrient	2001		2002		2003		2004		2005	
	Ca	Mg	Ca	Mg	Ca	Mg	Ca	Mg	Ca	Mg
Ca (R)	0.91*		0.99**		0.95*		ns		ns	
Ca (B)	0.94*		0.96**		0.99*		0.97*		0.93*	
Mg (R)		ns		0.79*		ns		ns		ns
Mg (B)		ns		0.92*		0.97*		0.81*		0.84*

R: in tree rows; B: between tree rows. **, * and ns: significant at $p < 0.01$, $p < 0.05$ and not significant, respectively. The values are means of four repetitions each year.

Table 1. Correlation coefficients between Ca and Mg concentrations in the soil (0–20 cm) between and in the tree rows and leaf Ca and Mg concentrations in guava trees over the experimental period

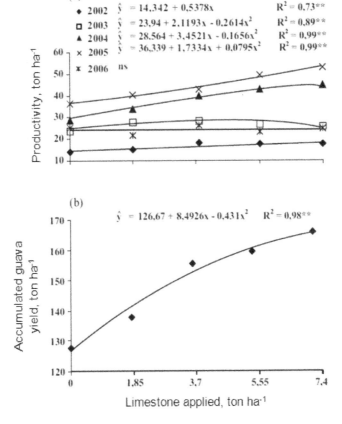

Figure 1. Effect of applying limestone on guava fruit yield for yearly harvests (a) and cumulated production (b).

Leaf Ca and Mg concentrations increased with lime rate and showed quadratic effects (Figure 2).

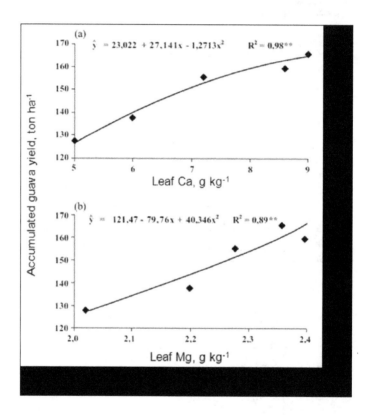

Figure 2. Relationships between the leaf concentrations of Ca (a) and Mg (b) and cumulated guava production.

The graph of the Ca/Mg ratio versus cumulated fruit yield shows, on the other hand, that the lower the ratio, the lower was fruit production (Figure 3).

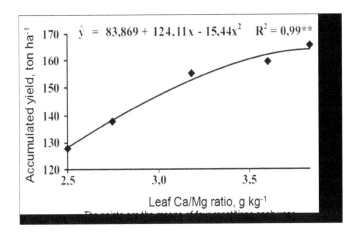

Figure 3. Relationship between leaf Ca/Mg and cumulated guava production for harvests cumulated between 2002 and 2006.

The cumulated fruit yield (2002-2006 harvests) increased quadratically with base saturation of the surface soil layer both in and between rows (Figure 4). Although model maximum goes beyond the values observed in the experiment, it can be inferred that satisfactory cumulated fruit production can be reached when V is closed to 50% in the rows and 65% between the rows.

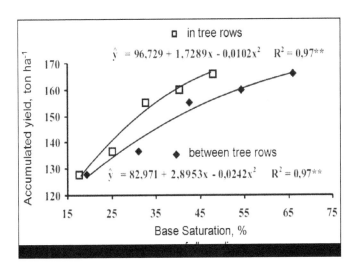

Figure 4. Relation between base saturation in the 0-20 cm soil layer, in and between the rows, and cumulated guava fruit production from 2002 to 2006.

The application of limestone to acidic soils promotes root development and, consequently, the uptake of water and nutrients. Determination of the exchangeable Ca and Mg concentration in the soil determined using an exchange resin gives an indication of the potential growth of the root system, especially at planting and tree formation stages and in situations where there are low levels of Ca. In [25], the authors evaluated the effects of liming on root system development and the mineral nutrition of guava trees grown in an acid dystrophic red latosol. They analyzed soil samples taken at four equidistant points 75 cm from the trunks in two layers (0–20 and 20–40 cm depth), in plots that received 0, 3.7 and 7.4 t ha^{-1} of limestone (reactivity = 94%). The corrective measure was applied before planting, incorporated with a disk plow in the 0–30 cm layer and harrowed to level soil surface. Forty-two months after incorporation of the limestone and the third year of guava tree cultivation, the roots were sampled with a cylindrical auger to assess the dry matter and lime content. Liming corrected soil acidity, increased base saturation and the availability and absorption of Ca by the plants and promoted root development. Calcium concentrations of 30 mmol$_c$ dm^{-3} in the soil and 7.5 g kg^{-1} in the roots were associated with greater root growth.

Furthermore, liming, by raising the amounts of Ca and Mg in the soil and the plant, can improve fruit quality. On this crucial point, [26] studied the effects of liming on the quality of guava fruits and observed that this practice did not affect their physical characteristics, such as weight, transverse diameter, length, flesh weight and flesh percentage. However, the application of limestone caused a linear increase in Ca concentrations in leaves and fruits, lowering weight loss of fresh matter and producing firmer fruits when Ca concentrations in the fruits reached at least 0.99 g kg^{-1}. Therefore, the provision for adequate Ca improved fruit quality regarding postharvest longevity.

These beneficial effects of Ca on fruit quality can be explained by the role this element in plant nutrition. In this respect, [27] observed that in guava fruits that received calcium in the form of limestone, cell walls and middle lamellae were well defined and structured, keeping the cells cemented. In contrast, in fruits from the plants not receiving limestone, the cell walls and middle lamellae were destructured and disorganized, respectively. The authors concluded that liming is an effective measure to improve the sub-cellular organization of guava fruits, contributing to tissue integrity.

Studies of the effects of liming on the biometric variables of plants are limited. In [28] the effect of limestone (CaO=45,6%; MgO=10,2% and reactivity=94%) application on trunk diameter, height and crown volume of Paluma trees was assessed under field conditions. The experimental design consisted of random blocks of five treatments and four repetitions. The treatments consisted of rising doses of liming in the 0-30 cm layer as follows: D_0 = no limestone; D_1 = half the dose to raise V to 70%; D_2 = full dose to raise V to 70%; D_3 = 1.5 times the dose to raise V to 70%; and D_4 = 2 times the dose to raise V to 70%. Field evaluations were carried out during seven years, starting at orchard's planting in 1999-2000 until the 2005-2006 growing season. The limestone increased trunk diameter and crown volume over the years. These results confirmed the importance of correcting soil acidity and the benefits of applying limestone on the biometric variables of guava trees.

Another popular fruit in Brazil is carambola (*Averrhoa carambola*) or star fruit that also responds well to soil acidity correction and fertilization. Investigating three-year old plants in field conditions, [25] found that the accumulation of root dry matter of this *Oxalidaceae* is boosted by limestone application, improving the uptake of nutrients and tree development.

Due to the low solubility of limestone, the best practice is to homogeneously incorporate this corrective material down to an adequate depth across the orchard area before planting the seedlings. Indeed, it is not recommended to till the soil after the trees have been planted nor is it advisable to apply limestone in the planting hole, especially along with phosphorus.

Despite the recognized importance of correcting acidity, there is virtually no study on the residual effect of this practice on carambola orchards. The only Brazilian work that assessed the effect of liming on carambola trees [20] was carried out in the country's main producing region between 1999 and 2006. The authors observed that the application of limestone produced significant changes in soil pH, potential acidity (H+Al), sum of exchangeable bases, base saturation and Ca and Mg concentrations at depths of 0-20, 20-30, 30-40 and 40-60 cm. Besides this, there were linear increases in pH, Ca, Mg, SB and V% and linear decrease of (H+Al) with limestone rates, both between and in the tree rows at all depths. The greatest changes occurred in the layer where the lime had been incorporated (0-30 cm) due to its low mobility.

Cumulated carambola fruit production from 2002 to 2006 as function of limestone application is shown in Figure 5. The base saturation between the rows remained higher than under the crown projection throughout the experiment. This behavior was expected because the management the trees required the application of high nitrogen doses which acidify the soil as a result of nitrification [5]. In addition, roots absorbed Ca and Mg and exuded H^+ [29]. Also, local irrigation was applied, which contributed to changes in the rates of ammonification, nitrification and denitrification. These rates, according to [30], are linked to the availability, location and forms of N in the soil profile. This is of fundamental importance in the mobility of nitrogen because of its low binding energy of nitrate to clay minerals and organic matter [31, 32], leading to nitrate leaching.

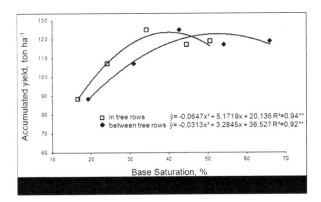

Figure 5. Relationships between cumulated fruit production from 2002 to 2006 and base saturation in the 0-20 cm soil layer between and within tree rows in a carambola orchard.

Figure 5 shows accrued fruit production (2002 to 2006 harvests) with rising base saturation in the 0-20 cm soil layer both between and within the rows. Therefore, 78 months after planting and acidity correction, maximum fruit production of 121 t ha^{-1} was obtained in the pH range between 4.6 and 5.0 where the base saturation reached 40% to 53% in and between the rows, respectively, and leaf Ca and Mg levels were 7.6 and 4.0 g.kg^{-1}, respectively [20].

Figure 6 shows differences in cumulated fruit production from 2002 to 2006. As expected, the fruit output rose with years, irrespective of limestone dosage. The reason is that the trees became more productive with the growth in height and leaf area as the study was conducted in a new orchard.

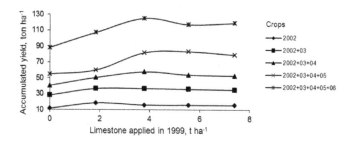

Figure 6. Cumulated production of carambola fruits related to limestone application rates at planting in 1999.

Figure 7 also shows that the cumulated fruit production increased yearly regardless of limestone rate. It is important to note that even after seven years, the control plots (zero limestone) still produced appreciable quantities of fruit, demonstrating the exceptional ability of the carambola tree to develop under adverse conditions.

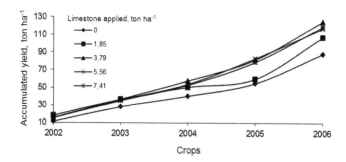

Figure 7. Cumulated carambola fruit production from 2002 to 2006 as function of limestone rates applied in 1999.

In a study of the economic aspects of liming, [22] observed that the cumulated production of carambola fruits as related with the application of different economically feasible rates of limestone coincided with the possible maximum output levels (Table 2). This occurred due to the high productive capacity of this species and the high average price of the fruit on the market.

In this study, we considered the price per ton of lime applied, divided by sales price per ton of carambola. The most economical dose was calculated based on the derived regression equation between the production of fruit and lime rates applied, making it equal to the exchange ratio, which was 0.05784.

Accumulated production	Economic dose	Increase in fruit yield	Cost of limestone	Profit	Production[1]
	t ha^{-1}	-------- t of fruit per ha -------			%
2002 to 2003	4.5	8.4	0.3	8.1	100
2002 to 2004	4.8	16.0	0.3	15.7	100
2002 to 2005	5.3	28.8	0.3	28.5	100
2002 to 2006	5.3	34.2	0.3	33.9	100

[1] Percentage of fruit production with the most economic dose in relation to maximum production.

Table 2. Economic dose of limestone as function of cumulated production of carambola fruits and limestone cost for the 2002-2006 period

The percentage of fruit production obtained with the most economic lime dosage in relation to the maximum production would be 100%. Therefore, the application of the economic dose allowed savings on limestone without significant loss of production. It is thus realistic to conclude that carambola trees respond positively from an economic standpoint to the application of limestone, which boosts fruit yield up to the dose considered adequate and recommended by [20].

The main carambola growing areas in Brazil are located in regions where soils are acid and show low base saturation which limits the normal development of plants hence orchard productivity. The effect of liming on trunk diameter, crown volume and height of carambola trees was evaluated in an experiment was conducted in the state of São Paulo in a red latosol (oxisol), in the period from August 1999 to July 2006 [33]. The limestone doses rates were 0, 1.85, 3.71, 5.56 and 7.41 t ha^{-1}. The soil was chemically analyzed and the three biometric variables were assessed during five consecutive harvests. The neutralization of the soil acidity provided an increase in the biometric variables during the entire experimental period. Liming increased trunk diameter, tree height and crown volume. The nutrients in limestone – Ca and Mg – positively influenced the development of the trees.

6. Liming of established orchards

The low solubility of most limestones limits the mobility of these materials in the soil profile, requiring initial incorporation to obtain a beneficial effect in the zone exploited by the roots.

In fruit orchards already in production, the procedure recommended by the official bulletins in Brazil is light incorporation of limestone in the tree rows [10]. However, it is probable that this recommendation would change if there were more research findings, considering the various phytosanitary problems that can occur due to lime incorporation, such as injuries and reduction of volume of the roots, with consequent risk of infections, dissemination of diseases in the orchard, increased attack by pests, especially mites, cochineals [34] and nematodes [35], as well as soil destructuring and compaction.

In orchards with adult trees, the application of limestone at the surface, without incorporation, will gradually neutralize the acidity below the surface due to the movement of the particles through the profile, at a rate of 1 to 2 cm a year, if moisture and drainage conditions are suitable [36]. Therefore, surface liming, even though possible, requires time to produce beneficial effects. However, the information mentioned above was obtained under edaphoclimatic conditions different than the tropics. Other studies have shown that it is possible to apply limestone at soil surface without incorporation and obtained satisfactory results over time.

To assess the effect of surface liming on soil fertility, plant nutrition and the productivity of guava trees, an experiment was conducted in a commercial orchard grown on a red-yellow Argisol (Ultisol) in the main guava producing region of the state of São Paulo [37]. The randomized block design was a 2 x 5 factorial scheme with three repetitions, where the factors were two types of limestone (common, with PRNT=80%; and calcined, with PRNT=131%), applied at five rates (0, 0.5, 1, 1.5 and 2 times the recommended rate to raise base saturation to 70%) without incorporation. The results showed that surface liming with either common or calcined limestone reduced soil acidity in proportion to lime rate down to depths of 0-10 and 10-20 in the established guava orchard. In the 10-20 cm layer acidity declined 6 to 12 months after applying calcined limestone and 24 months after the application of common limestone. The chemical composition of the leaves and fruits and fruit yield were not influenced by the lime treatments. The authors attributed this to the fact that because the trees are perennial, they need time to respond to change in management. They concluded that it is possible to surface lime established guava orchards to correct soil acidity both in surface and lower layers. However, there is a need for further research to determine the specific criteria for this crop and for adjusting the rates to the optimum base saturation for this type of liming.

Orange (*Citrus sinensis*) growing is another important activity in Brazil, occupying some 850 000 hectares. Brazil is the top producer of oranges accounting for 25% of world's total production and is the world's leading exporter of orange juice. This means that of each five cup of orange juice consumed in the world, three come from Brazilian orchards [38]. A field experiment was conducted in a grove of adult orange trees (Pêra variety) established on a red latosol (oxisol) with five rates of calcined limestone (PRNT=131%) applied onto soil surface without incorporation [39]. Treatment effects were monitored for three consecutive years on the movement of the lime through the profile 6, 12, 18, 24, 30 and 36 months after application, on the chemical properties of the soil, on plant nutritional status and on fruit yield. Surface application of calcined limestone altered the base saturation as well as the soil chemistry in the three successive soil layers (0-10, 10-20 and 20-40 cm). The maximum soil reaction upon liming occurred 12 to 18 months after lime application. Plants' nutritional status and fruit yield were significantly improved?. Cumulated production indicated that the ideal base saturation for orange trees was close to 50%. In the same experiment, [40] assessed the effect of limestone on the leaf Mn content. They found significant decreases in Mn levels with limestone additions. There was a high correlation between base saturation in the 10-20 cm soil layer and leaf Mn levels. Maximum fruit yields were associated with leaf concentrations between 33 and 70 mg Mn kg[-1]. Another liming experiment allowed determining, by the Compositional Nutrient Diagnosis (CND), the Diagnosis and Recommendation Integrated System (DRIS) and the mathematical chance methods, the adequate nutrient ranges to obtain high yield in Pêra orange groves [41].

Despite still insufficient scientific basis, practical experience showed that the pH, and consequently base saturation, should not be allowed to decline drastically in established orchards. The reason is that it is very hard to correct high acidity in the soil layers down to depths exploited by adult trees within a reasonable time interval. The best strategy in these situations is to apply small amounts of finely ground limestone annually (example.g, 1 ton ha[-1]), hence correcting acidity gradually, to avoid any sharp rise in acidity. It is clear, however, that soil analysis continues to be essential to assess the best timing and rates to apply lime at the surface without incorporation.

7. Conclusions

Soil acidity is a determinant factor limiting crop production in tropical areas. Fruit crops are perennials that exploit the same volume of soils during of time. As a result, soil acidity correction must be sustained in the roo zone to avoid aluminium toxicity and supply adequate amounts of calcium and magnesium to the crop. Research results show the economic advantages of liming to correct soil acidity and thus improve fruit yield and quality in Brazilian orchards.

Author details

William Natale[1], Danilo Eduardo Rozane[2], Serge-Étienne Parent[3] and Léon Etienne Parent[3]

*Address all correspondence to: natale@fcav.unesp.br

1 Unesp, Universidade Estadual Paulista, Campus Jaboticabal, Via de Acesso Paulo D. Donato Castelane, Jaboticabal, São Paulo, Brazil

2 Unesp, Universidade Estadual Paulista, Campus Registro, Registro, São Paulo, Brazil

3 ERSAM, Department of Soils and Agrifood Engineering, Université Laval, Quebec (Qc), Canada

References

[1] Ministerio da Agricultura Pecuaria e Abastecimento. MAPA: Estatisticas. http://www.agricultura.gov.br/vegetal/estatisticas (accessed 14 August 2011).

[2] Parent LE, Gagné G. Guide de référence em fertilization. 2nd ed. Quebec: Centre de référence en agriculture et agroalimentaire du Québec (CRAAQ); 2010.

[3] Instituto Brasileiro de Frutas. IBRAF: Fruticultura: síntese. http://www.ibraf.org.br/x-es/f-esta.html (accessed 11 March 2011).

[4] Sanchez PA, Salinas JG. Suelos acidos: estrategias para su manejo con bajos insumos en America Tropical. Bogotá: Sociedad Colombiana de la Ciencia del Suelo; 1983.

[5] Malavolta E. Manual de nutrição mineral de plantas. Piracicaba, Ceres; 2006.

[6] Mello, FAF. Origem, natureza e componentes da acidez do solo: critérios para calagem. In: MALAVOLTA, E. (ed.) Seminário Sobre Corretivos Agrícolas. Piracicaba: Fundação Estudos Agrários Luiz Queiroz; 1985. p.67-93.

[7] Ernani PR. Química do solo e disponibilidade de nutrientes. Lages: Udesc; 2008.

[8] Raij Bvan. Algumas reflexões sobre análise de solo para recomendação de adubação. In: RAIJ Bvan. (ed.) 20th Reunião Brasileira de Fertilidade do Solo e Nutrição de Plantas, 26-31 July 1992, Piracicaba, Brazil: Sociedade Brasileira de Ciência do Solo; 1992.

[9] Quaggio JA. Acidez e Calagem em solos tropicais. Campinas: Instituto Agronômico; 2000.

[10] Raij Bvan, Cantarella H, Quaggio JA, Furlani AMC. (eds.) Recomendações de adubação e calagem para o Estado de São Paulo. 2nd ed., Campinas: Instituto Agronômico; 1997.

[11] Natale W, Prado RM, Quaggio JA, Mattos Junior D. Guava. In: Crisòstomo LA, Naumov A, Johnston AE, (eds.) Fertilizing for High Yield and Quality Tropical Fruits of Brazil. 1st ed. Horgen/Switzerland: International Potash Institute; 2007. p103-122.

[12] Natale W, Prado RM, Rozane DE, Romualdo LM. Efeitos da calagem na fertilidade do solo e na nutrição e produtividade da goiabeira. Revista Brasileira de Ciência do Solo 2007;31(6) 1475-1485.

[13] Natale W, Rozane DE, Prado RM, Romualdo LM, Souza HA, Hernandes A. Viabilidade econômica do uso do calcário na implantação de pomar de goiabeiras. Ciência e Agrotecnologia 2010;34(3) 708-713.

[14] Pearson RW, Abruna F, Vice-Chances J. Effect of lime and nitrogen applications on downward movements of calcium and magnesium in two humid soils of Puerto Rico. Soil Science 1962; 93(2) 77-82.

[15] Harter RD, Naidu R. Role of metal-organic complexation in metal sorption by soils. Advances in Agronomy 1995;55(1) 219-263.

[16] Aoyama M. Fractionation of water-soluble organic substances formed during plant residue decomposition and high performance size exclusion chromatography of the fractions. Soil Science Plant Nutrition 1996;42(1) 21-30.

[17] Oliveira EL, Pavan MA. Control of soil acidity in notillage system for soybean production. Soil & Tillage Research 1996;38(1) 47- 57.

[18] Blevins RL, Thomas GW, Corneluis PL. Influence of no-tillage and nitrogen fertilization on certain soil properties after 5 years of continuous corn. Agronomy Journal 1977;69(3) 383-386.

[19] Marschner H. Mineral Nutrition in Higher Plants. London: Academic Press; 1995.

[20] Natale W, Prado RM, Rozane DE, Romualdo LM, Souza HA, Hernandes A. Resposta da caramboleira à calagem. Revista Brasileira de Fruticultura 2008;30(4) 1136-1145.

[21] Natale W, Coutinho ELM, Boaretto AE, Pereira FM. Goiabeira: calagem e adubação. Jaboticabal: Fundação de Apoio a Pesquisa, Ensino e Extensão; 1996.

[22] Natale W, Rozane DE, Prado RM, Romualdo LM, Souza HA, Hernandes A. Dose de calcário economicamente viável em pomar de caramboleiras. Revista Brasileira de Fruticultura 2011;33(4) 1294-1299.

[23] Silva SHMG, Natale W, Haitzmann-Santos EM, Bendini HN. Fert-Goiaba: Software para recomendação de calagem e adubação para goiabeira cultivar Paluma, irrigada e manejada com poda drástica. In: Natale W, Rozane DE, Souza HA, Amorim DA. (eds.) Cultura da Goiaba do plantio à comercialização. Jaboticabal: Universidade Estadual Paulista "Julio de Mesquita Filho"; 2009. p.281-284.

[24] Gros, A. Engrais: guide pratique de la fertilisation. 6th ed. Paris: Librairie de l'Académie d'Agriculture; 1974.

[25] Prado RM, Natale W. A calagem na nutrição e no desenvolvimento do sistema radical da caramboleira. Revista de Ciências Agroveterinárias 2004;3(1) 3-8.

[26] Prado RM, Natale W, Silva JAA. Liming and quality of guava fruit cultivated in Brasil. Scientia Horticulturae 2005;104(6) 91-102.

[27] Natale W, Prado RM, Môro FV. Alterações anatômicas induzidas pelo cálcio na parede celular de frutos de goiabeira. Pesquisa Agropecuária Brasiliera 2005;40(12) 1239-1242.

[28] Souza HA, Natale W, Prado RM, Rozane DE, Romualdo LM, Hernandes A. Efeito da Calagem sobre o crescimento de goiabeiras. Revista Ceres 2009;56(3) 336-341.

[29] Engels C, Marschner H. Plant uptake and utilization of nitrogen. In: Bacon PE. (ed.) Nitrogen fertilization in the environment. New York, Marcel Dekker; 1995. p.41-81.

[30] Miller AJ, Cramer MD. Root nitrogen acquisition and assimilation. Plant and Soil 2004;274(1) 1-36.

[31]   Reisenauer HM. Absorption and utilization of ammonium nitrogen by plants. In: Nielsen DR, Mcdonald JG. (eds.) Nitrogen in the Environment. London: Academic Press; 1978. p 157-170.

[32]   Lopes AS. Manual internacional de fertilidade do solo. 2nd ed. Piracicaba: POTAFÓS; 1998.

[33]   Hernandes A, Natale W, Prado RM, Rozane DE, Romualdo LM, Souza HA. Calagem no crescimento e desenvolvimento da caramboleira. Revista Brasileira de Ciências Agrárias 2010;5(1) 170-176.

[34]   Gravena S. Manejo integrado de pragas dos citros: adequação para manejo de solo. Laranja 1993;14(2) 401-419.

[35]   Silva GS. Manejo integrado de nematoides na cultura da goiaba. In: Natale W, Rozane DE, Souza HA, Amorim DA. (eds.) Cultura da Goiaba do plantio à comercialização. Jaboticabal: Universidade Estadual Paulista "Julio de Mesquita Filho"; 2009. p. 349-370.

[36]   Corrêa MCM, Natale W, Prado RM, Banzatto D. Liming to an adult guava tree orchard In: ARAGÃO FAS, ALVES RE. (eds.) 3rd International Simposium on Tropical and Subtropical Fruits, 12-17 September 2004, Fortaleza, Brazil: Internacional Society for Horticultural Science, Brazilian Society for Fruit Crop; 2004.

[37]   Lierop Wvan. Soil pH and lime requirement determination. In: Westerman RL. (ed.) Soil testing and plant analysis. 3rd ed. Madison: Soil Science Society of America; 1990. p.73-126.

[38]   Neves MF, Trombin VG, Milan P, Lopes FF, Cressoni F, Kalaki R. O retrato da citricultura Brasileira. Markestrat: Centro de Pesquisa e Projetos em Marketing e Estratégia; 2010.

[39]   Silva MAC, Natale W, Prado RM, Corrêa MCM, Stuchi ES, Andrioli I. Aplicaçao Superficial de calcário em pomar de laranjeira Pêra em pordução. Revista Brasileira de Fruticultura 2007;29(3) 606-612.

[40]   Silva MAC, Natale W, Prado RM, Chiba MK. Liming and Manganese Foliar Levels in Orange. Journal of Plant Nutrition 2009;32(4) 694-702.

[41]   Camacho MA, Silveira MV, Camargo RA, Natale W. Faixas normais de nutrientes pelos métodos ChM, Dris e CND e nível crítico pelo método de distribuição normal reduzida para laranjeira-Pera. Revista Brasileira de Ciência do Solo 2012;36(1) 193-200.

Chapter 7

Potassium Fertilization on Fruits Orchards: A Study Case from Brazil

Sarita Leonel and Luis Lessi dos Reis

Additional information is available at the end of the chapter

1. Introduction

What captures attention of those who are familiar with the production of fruits in Brazil is the diversity of species the country offers its people and foreign markets. From the temperate regions to the tropics and the equator line, only few varieties do not find their ideal climate and soil conditions across the country. It is common knowledge that brazilian people are privileged when it comes to the question of supply. Domestic consumers have year round acess to the types of fruit they prefer and, more recently, have even had the chance to select their product according to the different production systems. The planted area in 2010, it was 2.240 million hectares. The activity involves up wards of 5 million people throughout the country. In general, fruit growing is carried out on holdings of up to 10 hectares and provide enough income for the families to live life of quality.

2. The fig tree

The fig tree (*Ficus carica*, L.) originated from Asia Minor and Syria, in the Mediterranean region, and was first cultured and selected by Arabs and Jews in Southwest Asia. It is one of the oldest plants cultivated in the world – since prehistoric times – and is considered by ancient people as a symbol of honor and fertility. According to botanists from the American University Harvard, Middle Eastern fig trees were the first species cultivated by humans, 11,400 years ago. Researchers have found the remains of small figs and dry seeds buried at a village in the Jordan Valley located to the north of Jericho. The fruits were well conserved, which indicates they were dried for consumption [1].

The fig is one of the most popular food that has been sustaining humanity since the beginning of History. The fruit was used to feed advanced Olympic athletes and was offered to the winner as the first Olympic medal. The tree was described in many passages from the Bible as sacred and respected by man. During the period of the great discoveries, the fig was disseminated to the Americas. In Brazil, the fig tree was probably introduced by the first colonizing expedition in 1532 in São Paulo State.

In Brazil, economic exploration of the fig tree only started from 1910, when it was first commercially cultivated in Valinhos region, São Paulo State, where crops are restricted to only one cultivar – 'Roxo de Valinhos'. This cultivar was from a region close to the Adriatic Sea in Italy and was introduced in Brazil, in the region of Valinhos, at the beginning of the 20th century by the Italian Lino Bussato.

'Roxo de Valinhos' fig plant is vigorous, productive and adapted to the drastic pruning system; this practice was adopted to help control pests and diseases. This is the only cultivar that has economic value due to its rusticity, vigor, and productivity; in addition, it is a product sensitive to handling and easily perishable. Production can be directed to industry for the fabrication of green fig compote, jam and crystallized fig, or for consumption of raw fruits.

The fig tree is commercially cultivated in the Brazilian states Rio Grande do Sul (39.42%), São Paulo (35.15%), and Minas Gerais (18.75%). In São Paulo State, the production is mainly destined for the market of raw fruits, whereas in the other states it is directed to industrial processing. According to data from the Brazilian Ministry of Agriculture (2008), Brazil produced 26,476 t figs in 2006, in a 3,020ha area, resulting in an average national productivity of 8.8 t/ha.

The culture is interesting for Brazil as it may lead Brazilian exportations to be incorporated between harvests in Turkey, which is the world's main producer of figs. Brazil is a great furnisher of figs to the world; 20 to 30% of the total volume produced in the country is destined for exportation. Commercialization is done in boxes containing 1.6 Kg of the fruit [1].

2.1. Potassium fertilization in fig orchards

Little is known about the nutritional demands for the fig tree culture. The results available mostly discuss the use of organic fertilizers, where those appear as favorable practices, both in the development and the production of fig trees. Experiments with different sources and doses of nitrogen had also been widely performed, however, little is known about the demands of the other nutrients. According to [2], balanced and satisfactory mineral nutrition factors during the phase of formation of the plants assure good crops in the production phase of the plant.

Thus, in the absence of systematic studies for this purpose, the fertilizations of this fruit tree are performed mostly in an empiric way, mainly during implantation and formation of the trees. Likewise, nutritional diagnosis of plants through foliar analysis, although being a widely recognized valuable instrument for perennial plants, is incipient in the case of fig culture, often with conflicting values and absence in case of diagnosis with use of petioles.

Although the nutritional demands of the fig tree are of fairly knowledge, its measurement involves components of a very complex range, since the nutrient demands are closely related to the aspects of the species' physiology. During reproductive phase, the nutritional requirements have a component which is easily measurable and highly important in the evaluation of nutritional demands, the export of nutrients within fruit crops. However, during plant formation phase, the nutritional demands become difficult to determine since those are only for the growth and establishment of the plant, as well as the analyses of development the plants are rarely done in this period. In this phase excessive fertilization is performed according to visual diagnosis done by the producer, which is not uncommon.

2.2. Nutritional diagnosis and fertilization recommendations

According to [3], the knowledge of any needs or excesses of chemical elements responsible for the metabolism of plants and, due to the vegetation and productivity of fruit trees, it constitutes a necessary and indispensable step for corrective measures, since the fertilizing recommendations consist in the employment of fertilizer amounts, aiming to correct the element or limiting factor detected by the diagnosis.

The fertilizing recommendations during the formation period of the fig trees had been advocated exclusively from interpretations of soil analysis. In case of planting fertilizations, the recommendations are made by subjective criteria, not taking into consideration the content in the soil [4]. However, according to [3,5], soil analyses may be used to follow up the fertility of the soil and fertilizing recommendations during the development of the plants, because when used concomitantly with diagnosis methods may yield better results. The nutritional state of the plant can reveal the availability of nutrients of the soil and the ability the plant has to absorb them. Yet, fertilizing recommendations based on nutrient demands for fruit production, growth of branches, trunk and roots, during the phase of plant formation, cannot be considered a practice sufficiently broad, since such requirements are hard to measure. According to [6,7], the nutritional demands are better evaluated for plants at full production, where the crops of unripe and ripe fruits constitute the main sources of nutrient extracting sources.

2.3. Effect of potassium fertilizer on the fig tree

Due to scarce information on fertilization and nutrition of the fig tree, coupled to the evaluation of its effects on the nutritional state, a research was conducted using the levels of potassium fertilization, during the period of plant formation [8].

2.4. Methodology

The experiment was carried through in field conditions at the Orchard of Experimental Farm, of São Paulo State University, Faculty of Agronomic Sciences, Campus of Botucatu, São Paulo, Brazil, located at 22^0 $51'$ $55''$ South Latitude, 48^0 $26'$ $22''$ Western Longitude, with altitude of 830 meters. The predominant climatic type at the location, according to [9,10], based on the KOEPPEN international System, is included in the Cfb, namely the temperate climate without dry winter, mean temperature of the coolest months below 18°C and the ones from the warmer months below 22°C, with annual mean precipitation of 1314 mm, reaching in the driest month (August), a 26 mm average. The climate conditions observed during the conduction of the experiment are in Figures 1 and 2.

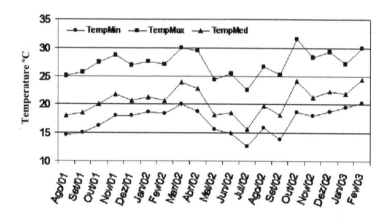

Figure 1. Maximum, mean and minimum temperatures observed during the conduction of the experiment. UNESP/ Botucatu - SP, Brazil, 2012.[8].

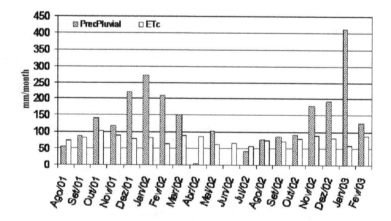

Figure 2. Pluviometric precipitations and evapotranspiration of the fig tree culture during the conduction of the experiment. UNESP/Botucatu - SP, Brazil, 2012.[8]. Source: Evapotranspiration of Class A Tank (ECA) - Area of Environmental Sciences (FCA). Evapotranspiration from the Culture (ETc) – Calculated by Culture Coefficient (Kc) data.

The soil is Rhodolic Haplo Udalf, according to the criteria established by [11]. The results of soil analysis of the 0-20 cm layer performed before and after saturation increasing by basic cations, according to the methodology in [12], are presented in Tables 1 and 2, respectively.

pH CaCl$_2$	MO g dm^{-3}	P resin mg dm^{-3}	H + Al	K	Ca	Mg	SB	CTC	V %
						mmol$_c$dm^{-3}			
4.2	24.0	3.0	77.0	1.5	12.0	5.0	19.0	96.0	19.0

Source: Soil Fertility Laboratory – Department of Environmental Resources – Area of Soil Science.

Table 1. Chemical characteristics of the soil where the experiment was performed before saturation increasing by bases. UNESP/Botucatu-SP, Brazil, 2012. [8].

pH CaCl$_2$	MO gdm^{-3}	P resin mgdm^{-3}	H + Al	K	Ca	Mg	SB	CTC	V %
						mmol$_c$dm^{-3}			
5.6	31.0	14.0	32.0	1.3	37.0	21.0	60.0	91.0	66.0

Source: Soil Fertility Laboratory – Department of Environmental Resources – Area of Soil Science.

Table 2. Chemical characteristics of the soil after saturation increasing by bases and planting fertilization. UNESP/Botucatu-SP, Brazil, 2012. BRIZOLA et al. (2005)[8].

The experiment was performed adopting the randomized block design, in an experimental scheme of subdivided parcels along the time, with four repetitions. The parcels were composed by potassium levels, sub-parcels by years and sub-parcels by harvesting months. The experimental unit was composed by three useful plants of the fig tree from cv 'Roxo de Valinhos', completely surrounded by border plants, in 3 x 2m spacings among plants and among lines, thus composing an useful area of 18m^2 for each experimental unit.

The main treatments, potassium fertilization levels (Table 3), were administered in the period from August to September of the agricultural cycles using increasing doses in arithmetic progression, in which the levels of the second cycle were equal to the first.

Treatments	Potassium Levels
K 0 (Witness)	Zero K$_2$O
K I	30g K$_2$O plant^{-1}
K II	60g K$_2$O plant^{-1}
K III	90g K$_2$O plant^{-1}
K IV	120g K$_2$O plant^{-1}
K V	150g K$_2$O plant^{-1}

Table 3. K$_2$O levels applied in the experiment. UNESP/Botucatu-SP, Brazil, 2012. [8].

Potassium fertilizations began from seedling fixation, potassium chloride used as a nutrient supplier, the levels during the first year adopted according to the recommendation in [4], with two levels lower and three levels higher than the 60 g K_2O/plant recommendation. For levels higher than 60g K_2O plant^{-1}, the applications were divided in three occasions, with 20-day intervals. Nitrogenized fertilizations were also used using ammonium sulphate in four applications, placing 15g nitrogen plant^{-1} at each occasion. The fertilizations were applied in the projection of the crown of the tree and superficially incorporated using a shovel in the two years of conduction of the experiment. The use of phosphorus was done only during the plantation, at the amount of 100g plant^{-1} of P_2O_5, applying simple superphosphate.

The evaluation of the nutritional state of fig tree plants was performed through the diagnosis of the leaf and petiole, in three months within each evaluation year: October, December and February. The analyses of macronutrient content and branches and fruit accumulations were performed to evaluate the extraction of nutrients by the fig tree. The evaluations were obtained during the growth and plant production periods, the collections performed in three periods (October, December and February), evaluating: number of leaves, length of branches (cm), trunk diameter, dry matter of branches and production of fruits.

2.5. Results

The results obtained evidenced that the content of macronutrients in branches were not influenced by potassium fertilization in the crown. According to [13,14], the interactions between ions assume the existence of a certain relationship within those in soil solution (nutrient availability), this relationship being able to manifest itself in the form of nutritional imbalance, where the leaves are the first organs to manifest those changes, both at the level of contents and visual symptoms. Regarding the macronutrient content in branches, those were not influenced by potassium fertilizations. Thus, it can be accepted that such interactions in levels of content in branches are observed in more prolonged conditions of nutritional imbalance.

The growth of branches and the number of leaves by branch increased with fertilization and, accordingly, positive responses were obtained with potassium fertilization in the production of dry matter of branches and fruits (Figures 3 and 4).

According to [13], potassium deficiencies may reduce the photosynthetic activity and increase respiration, reducing the supply of carbohydrates and with consequent effects on the growth of the plants. For [15], the physiological functions played by potassium are directly involved in protein synthesis, in the use of water and in the translocation of carbohydrates, conditions which, when perfectly functional, may lead to plant growth.

The evaluation of structures of the plant showed that the leaves were the organs that presented the highest levels of nitrogen, phosphorus, calcium and magnesium, while the fruits were the organs that presented the lowest levels of macronutrients (Table 4).

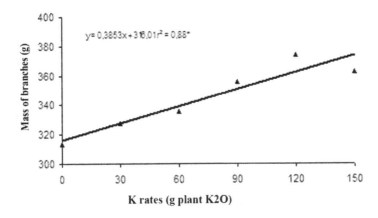

Figure 3. Effects of potassium fertilization in the dry matter of branches of the fig tree. FCA/UNESP/Botucatu, SP, Brazil. [8].

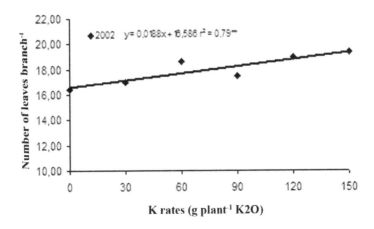

Figure 4. Effects of potassium fertilization in the number of leaves by fig tree branch. FCA/UNESP. Botucatu, SP, Brazil. 2012. [8].

Comparing the results found for foliar contents in the experiment to those suggested by [16] as optimal for well-nourished plants, it is observed that only nitrogen and potassium presented levels lower than those found by the authors, whereas in such concentrations, according to the same authors, those nutrients were already approaching deficiency zone. In [17], foliar contents indicated as satisfactory for fig tree culture are in ranges of 22-24 for N; 1.2-1.6 for P; 12-17 for K; 26-34 for Ca and 6-8 g Kg^{-1} for Mg, whereas, in comparisons, only for Ca and Mg content lower than those indicated by the authors were detected. In comparison to the values indicated by [18] for foliar contents, calcium and magnesium presented values lower than those considered optimal for the culture, however, there was no perception of any manifestations of nutritional deficiency symptoms connected to those two nutrients, even in the treatment where the highest doses of potassium were applied.

Nutrient (g Kg^{-1})	Leaves	Petioles	Branches	Fruits
Nitrogen	25.57 A	11.14 B	10.36 B	7.995 C
Phosphorus	2.096 A	1.475 B	1.033 BC	0.777 C
Potassium	21.89 B	31.82 A	2.213 D	8.620 C
Calcium	19.25 A	10.75 B	6.982 C	1.863 D
Magnesium	5.675 A	4.262 A	1.981 B	0.727 B
Sulfur	1.707 B	3.064 A	0.960 B	0.766 B

Means followed by the same letter, in the same line, are not significantly different from the means by contrast, at the level of 5% of likelihood by F test.

Table 4. Mean content of macronutrients in leaves, petioles, branches and fruits of fig tree undergoing six levels of potassium in top-dressing fertilization. UNESP/Botucatu-SP. [8].

Regarding the contrasts of means between the content in leaves and petioles, it was noticed that the potassium and sulfur content were lower in leaves, whereas for magnesium, the contents were not different in leaves and petioles. For nitrogen, phosphorus and calcium the foliar contents were higher than those found in petioles, results in agreement with those found by [16], where contents of N of 33.9 and 15.1; P of 2.0 and 1.6; K of 26.8 and 45.9; Ca of 16.7 and 11.9; Mg of 6.3 and 8.4; and S of 2.0 and 4.4 g Kg^{-1} were found in leaves petioles, respectively.

It was also observed that macronutrient content in leaves presented good correlations to those determined in petioles, and the petioles had better correlation coefficients with dry matter production and fruit production, making them preferential for the analysis of nutritional status of fig trees being formed (Table 5). Such results are in agreement with literature data, which indicate that the petiole is the most appropriate organ for evaluation of potassium in the plant [16,17,19].

Ratios Between Nutrients	Correlation Coefficient (r)		Significance test (F)	
	First crop cycle	Second crop cycle	2001/2002	2002/2003
N (leaf x petiole)	0.738	0.806	0.009**	0.000**
P (leaf x petiole)	0.591	0.634	0.040*	0.025*
K (leaf x petiole)	0.715	0.761	0.001**	0.002**
Ca (leaf x petiole)	0.771	0.829	0.003**	0.000**
Mg (leaf x petiole)	0.612	0.651	0.034*	0.018*
S (leaf x petiole)	0.658	0.660	0.018*	0.019*
K (leaf x soil)	0.386	0.773	0.215ns	0.003**
K (petiole x soil)	0.417	0.736	0.176ns	0.009**

ns = Non-significant a P>5% by F test; * Significant at 5% of likelihood; * * Significant at 1% of likelihood.

Table 5. Correlations of macronutrient content in leaves and petioles of fig tree undergoing six levels of potassium in top-dressing fertilization. UNESP/Botucatu-SP, Brazil, 2012. [8]

The results of fruit production (Figure 5) show that increases in potassium levels in top-dressing increased linearly with the production; however, the trend of the equation indicates an adjustment for a cubic equation when using higher levels of K_2O. Thus, the availabilities of potassium above 90 g K_2O plant^{-1} could be considered as luxury consumption, since those would not be increasing the production values.

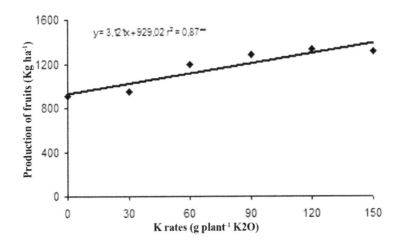

Figure 5. Effects of potassium fertilization in total production of fruits of developing fig tree. FCA/UNESP/Botucatu, SP, Brazil. 2012. [8].

In [20] no effects were obtained for the higher doses of potassium, although the employment of the dose of 60g K_2O plant[-1] had been about 40% higher than the dose of 30g K_2O plant[-1]. The authors justified such results due to the high variation coefficient obtained for the analysis of harvesting of unripe fruits. For [3], the effects of potassium fertilizations on fruit trees are more conditioned to aspects of quality than quantity, since this element is not in limiting amounts for the development of the plant.

2.6. Conclusions

The results showed that potassium fertilizations provide increases of production of dry matter of branches and fruits, where better results were associated with levels of 90 g K20 plant[-1], in a stand of 1600 plants per hectare and in soils under conditions of low and medium fertility in potassium.

3. The apple tree

The apple tree (*Malus domestica*) origin center is the Caucasus region, in the Asian montains and in the East of China. It is supposed that the development of the casual species have been initiated 20.000 years ago. It seems that the Greeks in the classical ancient times had cultivated apple tree, in fact in the roman empire the apple tree culture was already widespread. In Brazil, the beginning of the apple tree culture occurred probably in Valinhos municipality, state of São Paulo in 1926 [21].

Apple is among the fourth most consumed fruits in the world. In Brazil it is commercialized during the twelve months of the year and distributed all over the country. Except its consumed *in natura* it is utilized in puree, jam, dry fruit, concentrated juice and fermented beverages. The apple tree fruit is rich in peptic substances and cellulose that together with lignin constitute fibers [21].

The apple orchards in Brazil initiated in the end of the 60's and beginning of the 70's. Since this date, Brazil depended on importation to supply the apple market. But trough the government supporting, now a days, the evolution of the crops was fast, getting to 34 thousand of hectares and a production of about 850 thousand ton, concentrated in the Santa Catarina state. From 1988 Brazil started to export apples reaching self-sufficiency in 1998 when the exportations exceed the importations [21].

3.1. Potassium fertilization in apple orchards

Potassium (K) is the most extracted nutrients from the soil by apple fruits. Currently, with the use of new technologies, yield may reach values higher than 100 t ha[-1] [22]. Which increases the nutrients demanded by the apple.

The effective fruiting of apple trees is influenced by N, which plays an important effect in floral bud formation and increases the period when the ovule can be fertilized [23]. K increases sugar translocation to sink tissues, promoting their growth [24]. Thus, fruits from K-deficient plants have reduced size [25,26,27,28]. Which can reduce overall yield [22].

According with [29,30], an excess of K can affect the calcium (Ca) nutrition, increasing the intensity of physiological disorders related this nutrient, including the bitter pit, cork spot and lenticel blotch pit, among others. Increasing K as well as N rates can decrease flesh firmness, reducing the storage life of apples.

Fertilizer recommendations for apple in Santa Catarina (SC) and Rio Grande do Sul (RS) states Brazil are based on soil and leaves chemical analysis, shoot growth and orchard productivity [31]. The amount recommended for each year varies from 0 to 100 kg ha^{-1}. These recomenadtion were obtained from results of research conducted in Fraiburgo/SC and Vacaria/RS, or adapted from other production regions around the world. Reginal fertilization test are quite important to São Joaquim/SC, considering that this region presents very stony and shallow Inceptisols and the mean temperatures are lower when in comparison to other production regions in Brazil.

3.2. Methodology

In [32] made an research with the objective to evaluate the effects of long-term annual additions of K to the soil on yield, fruit size, mineral composition and Ca-related disorders of 'Fugi' apples for São Joaquim, Santa Catarina state, Southerm Brazil (28° 17' S, 49° 55' W). The experiment was conducted in the growing seasons from 1998 to 2006 in three commercial orchards of 12, 16 and 19 years old. Clay content and chemical characteristics of the soil from the experiment orchards, at the beginning of the experiment, are presented in Table 6.

The experimental plots comprised five plants, spaced 4.5 m between plants by 6.0 m between rows in one orchard and 3.0 by 6.0m in the other two, with the three central plants used as measurement plants. Trees were trained on a central leader system and received the same pruning and thinning practices as recommended for apple commercial orchards.

Attribute	Orchard 1	Orchard 2	Orchard 3
pH (H$_2$O)	6.8	6.4	6.6
P (mg dm^{-3})	33.0	45.0	63.0
K (mg dm^{-3})	141.0	240.0	258.0
Ca (mmolc dm^{-3})	89.0	112.0	119.0
Mg (mmolc dm^{-3})	60.0	62.0	64.0
Organic matter (g dm^{-3})	50.0	49.0	65.0
Clay 9 g dm^{-3})	300.0	380.0	300.0

Table 6. Soil testing results before experiment implementation (1998). [32].

3.3. Results

The results showed the apple yield was increased by K fertilization in four of eight evaluating growing seasons (Table 7), corroborating the results obtained in a long term experiments in south Brazil. The maximum increment in yield due to K fertilization ranged from 8.4 t ha^{-1} to 17.5 t ha^{-1}, representing increases of 16,0% and 68,3% in fruit yield, respectively, as compared to trees not receiving K in these years.

In the first and third year no effect of K fertilization on yield was detected, because of the high exchangeable K content in the soil in all orchards prior to establishment of the experiment (Table 6).

Yield was more consistently increased by K fertilization after the 2002/2003 growing season, when exchangeable K contents were reduced in the plots without fertilization. The absence of response in the 2005/2006 growing season can be attributed to the increase in K levels of the plant, as a result of lower yields observed in the previous two growing seasons (Table 7).

K$_2$O	Growing Season								
	98/99	99/00	00/01	01/02	02/03	03/04	04/05	05/06	98-06
Kg ha^{-1}	--t ha^{-1}--								
0	50.7*	52.5	38.7	41.0	35.3	28.3	25.6	46.6	318.7
50	50.5	56.8	38.9	43.3	40.6	31.1	38.8	55.3	355.3
100	50.8	60.9	40.7	47.5	46.4	36.2	42.4	54.7	379.6
200	48.7	56	44.1	45.6	47.7	39.7	43.1	49.2	374.1
Mean	50.2	56.5	40.6	44.3	42.5	33.8	37.5	51.4	
Coefficient Variation	34.5	21.1	35.9	29.7	17.8	24.6	20.5	31.3	

* Average values (n = 12)

Table 7. Average annual and cumulative fruit yield (1998-2006) for 'Fugi', as affected by annual surface adition of K. [32].

3.4. Conclusions

Yield size of apple were influencied, in a non interactive way, by K fertilization. Depending upon the growing season, yield and size of the fruit were often increased in response to annual addition K to soil, with fruit size more affected by K.

4. The pineapple in Brazil

The pineapple tree *Ananas comosus* (L.) Merril belongs to the Bromeliaceae family, subclass of subclass of Monocotyledonous and gender *Ananas*. It is a plant native to South America, covering latitude from 15° N to 30° S and longitude from 40° E to 60° W. About 50 genders and 2000 species of Bromeliaceae are known, some of them showing high ornamental value and others producing fibers excellent for cordage [33].

According to [34], the fruit of the pineapple tree is composite or multiple types called syncarp or sorosis formed by the coalescence of individual fruits, berry type, in a spiral on the central axis which is the continuation of the peduncle. The fruit is parthenocarpic, i.e., formed without the advent of fecundation. This fecundation may be possible but generally the varieties cultivated are self-sterile. According to [35], the skin of the fruit is composed of sepals and tissues of bracts and apices of the ovaries, while its edible portion consists mainly of the ovaries and bases of the sepals and bracts, as well as the cortex of the central axis.

The leaves of the pineapple tree, which can reach a maximum 70 to 80 per plant, are rigid and serous in the surface and protected by a layer of hair (trichomes) found in the lower surface, which reduces transpiration to a minimum [36]. The leaves are inserted in the stem and arranged in a rosette where older leaves are located on the outside of the plant and the newest in the center [37]. The "D Leaves" are the newest among the adults and the most physiologically active within all leaves, the reason why they are used in evaluations of nutritional status of the plant and in measures of growth [36].

The radicular system of a mature plant is of the fasciculated type and is located in the superficial part of soil surface. The majority of the roots are located in the first 15 to 20 cm of depth. The process of flowering begins with the reduction in vegetative growth velocity with a corresponding increase in collection of starch in leaves and stem [38].

The pineapple, native to Brazil, thrives under the Country´s ideal soil and climate conditions, where it is grown from North to South, and its economic importance is acknowledged everywhere.

Pearl is the major variety in Brazil while in the world the Smooth Cayenne variety is the most popular. Although having an acid taste, this variety boasts the characteristics required by the consumers. To please consumers' eyes and palate, pineapples must have yellow pulp and skin, cylindrical shape, small crown and a taste similar to the Pearl variety, in addition to normal packaging and labeling requirements.

Brazil is one of the world greatest growers of pineapple producing around 2.5 million tons in 2008 [39]. Despite the importance of potassium fertilization for this crop, there is a lack of information about the effects of different sources of K on fruit yield.

The pineapple tree is considered the worldwide third most cultivated fruit tree and exhibits a market which annually moves about US$ 1 billion dollars, being cultivated in more than 50 countries [39]. The Philippines followed by Thailand are the world biggest producers of pineapple with an annual production of two million tons, next in 6th place Brazil reaches around 1.47 billion fruits per year and, in the sequence, India, Nigeria and México [40].

In Brazil the pineapple is traditionally cultivated under rainfed conditions, in sandy, acid and low-fertility soils, with limitations for Ca, Mg and K and unbalances on the ratios among those cations [41]. In real values potassium and nitrogen are the most absorbed elements by the pineapple tree. The size and weight of the fruit are variables directly related to nitrogen, while potassium is linked to the physical-chemical quality of the fruits [42,43,44,45].

4.1. Potassium fertilization in pineapple

The nutrients required the most by the pineapple tree and which influence its growth are potassium and nitrogen [46]. Potassium is the nutrient which accumulates the most in the plant, markedly interferes in product quality and also in culture productivity; nitrogen mostly influences the fruit mass. The pineapple tree is not very demanding in phosphorus and its importance to the plant is mainly in floral differentiation and fruit development phase [47]. In [48] it is mentioned that an increase of N reduces the acidity of the fruits, but it can or cannot decrease soluble solids. According to [49,50], the extraction for macronutrients in decreasing order is expresser: K, N, Ca, Mg, S and P and for micronutrients: Mn, Fe, Zn, B, Cu, Mo.

Potassium is an important enzyme activator, responsible for opening and closing of stomata and carbohydrate transportation. It increases the content of soluble solids and acidity, improves the color and firmness of the skin and pulp and increases the mean weight and diameter of the fruit, and also decreases the emergence of internal darkening of the fruit [51].

According to [52] potassium fertilization can be supplied with potassium chloride, potassium sulphate, potassium and magnesium double sulphate and potassium nitrate, the two last ones being harder to find on the market and more expensive. The minimum guarantees and characteristics are presented in Table 8.

Source	Minimum guarantees/characteristics
Potassium chloride KCl	58% K_2O. Potassium in the form of chloride determined as K_2O soluble in water.
Potassium sulphate K_2SO_4	48% K_2O and 15% de S. Potassium in the form of sulphate, determined as K_2O soluble in water.
Potassium nitrate KNO_3	44% K_2O and 12% de N. Potassium determined as K_2O soluble in water. Nitrogen in the nitric form.
Potassium and magnesium double sulphate $K_2SO_4 . 2MgSO_4$	20% K_2O, 10% de Mg and 20% de S. Potassium and magnesium determined as K_2O and Mg soluble in water.

'Source: Instrução normativa n° 5 do Ministério da Agricultura, Pecuária e Abastecimento [53].

Table 8. Sources of potassium used in pineapple trees[1].

The mostly used source by the producers is potassium chloride due to its low cost, but its composition has chloride which is a toxic element. The combination of chloride from the fertilizer with the one present in the irrigation water in the region, further increases the toxicity of this element on the plant. Potassium sulphate is the most appropriate source, for being less harmful to the crop.

The presence of chlorine affects the starch and sugar contents in the plant. High concentrations may prevent fructification and potassium absorption, reducing the size of the fruit, the sugar and starch contents, increasing the acidity, symptoms similar to K deficiency [54].

When evaluating different combinations of potassium sulphate and chloride, supplied in pits of basal leaves, at 30, 90, 180 and 270 days after planting at the dose of 8 g plant^{-1} K$_2$O, [55] no differences were observed in the production and quality of the fruits when cultivating Smooth Cayenne, although a trend of slight increase in acidity and decrease in total soluble solids/total titratable acidity ratio had occurred to the extent that the applications of sulphate were replaced by potassium chloride. In addition, no visible registered symptoms of foliar burning were registered by the use of KCl nor changes in fruit color.

In [56], the effect of fertilization (potassium sulphate, 0; 8 and 16 g plant^{-1} K$_2$O) on the production and quality of the fruits from Smooth Cayenne cultivar, in Argissolo Vermelho dos Tabuleiros Costeiros de Pernambuco, containing 17 mg dm^{-3} K was evaluated. Significant effects of K on the content of soluble solids were seen, which reached maximum values at the dose of 15.6 g plant^{-1} K$_2$O.

When evaluating the effect of four doses of K (0, 413, 722 and 1.031 kg ha^{-1} K$_2$O), applied in the form of potassium chloride, in low fertility soil from Minas Gerais, [57] observed better use of K by the plants in the presence of liming and that the doses of K$_2$O which maximized the production were greater in more elevated doses of N (236 and 720 kg ha^{-1} K$_2$O to 10 and 15 g plant^{-1} de N, respectively). The increase in doses of K increased the foliar content of K and reduced the Ca and Mg content. It also increased the content of total soluble solids and fruit acidity, granting good balance in the SST/ATT, *ratio*.

Overall, the doses of K to maximize the quality attributes of the fruits are greater than those to maximize the production. In this context, [44] when evaluating the doses of K (0, 175, 350 and 700 kg ha^{-1} K$_2$O) necessary to obtain maximum physical and quality yield of pineapple fruits Smooth Cayenne cv. observed that the doses of K positively influenced the size of the fruits and a total production in addition to increase the content of vitamin C, soluble solids and total acidity. However, the dose of K to maximize the size of the fruits (569 kg ha^{-1} K$_2$O), was higher than the one to maximize the production (498 kg ha^{-1} K$_2$O).

Due to a long cycle culture and high K demand throughout the cycle, the application of potassium fertilizers in the pineapple tree should be divided to meet the demands of the culture, minimize losses, increase efficiency of fertilizations and improve fruit quality [43].

In Brazil, the main soil classes are Latosoil and Argisoil with elevated degree of intemperization and little presence of potassium minerals. In less intemperized soils, like Neosoils, Vertisoils, Luvisoils and Chernosoils, more rare in the Country, there are larger quantities of potassium minerals, like feldspates and mica, which may represent important sources of the nutrient [58]. Thus, soil contents maintenance appropriate to plants becomes extremely important in cultivation of the pineapple tree.

4.2. Sampling and chemical analysis of the soil

Some technical criteria should be adopted in soil sampling, since failure in the collection of soil samples generate errors that cannot be corrected later by soil analysis. All care should be taken in order to the samples being representative of the areas to be cultivated.

The sample collection should be performed before plantation, with enough time for the corrective to have time to react and to perform the fertilization step. The area to be sampled should be divided in homogeneous plots. For this division, observe the topography, vegetable covering, area history, drainage, soil texture, soil color and further related factors.

At the samples withdrawal the arable layer, which normally is more intensely changed by plowing, harrowing, correctives, fertilizers and culture residues, is considered. Therefore, sampling should be performed in this layer from 0 to 20 cm depth. For the analysis of sub superficial acidity and availability of sulfur the depth from 20 to 40 cm should be collected.

For larger representativeness, 15 to 20 single samples should be collected, using an instrument which provides equal volume between collections, at randomly distributed points in each area; the set of single samples will constitute the composite sample (500g homogenized fraction).

The composite sample from each area should be forwarded to a lab for soil chemical analysis for fertility purposes that present performance control of its results by IAC, easily identified by the seal. The requested analyses should be the basic (pH, MO, P, H+Al, K, Ca and Mg) and micronutrients (B, Cu, Fe, Mn and Zn) ones. Optionally, the analyses of Al, SO_4^{2-} and texture can be requested, as indicated.

4.3. Evaluation of nutritional availability

Foliar diagnosis was performed for plants status nutritional evaluation. Foliar analysis allow monitoring the of fertilizers used, however, is necessary caution for sample collection, working within sampling standards and criteria.

4.4. Foliar diagnosis

In leaf sampling, it is important to establish criteria to define the plots, grouping plots with similar characteristics regarding cultivated variety, age, phenology, handling, productivity and which ones belong to areas with homogeneous soils.

For the pineapple tree it is recommended to collect the Leaf "D" (Figure 6), considered metabolically more active, which is the last well-developed leaf, generally the longest one, forming, in general, a 45^0 angle relative to the soil (Figure 7). The sample collection should be performed by the period of floral induction [59]. However, the sampling time indicated for the pineapple tree does not allow corrections in the current crop. The suggestion is to collect at least 25 leaves of different plants, from each uniform plot, randomly taken, considering one leaf per plant [17].

Figure 6. Overview of a pineapple plant with indication of the leaf to be sampled for chemical analysis (Leaf "D"). Photo: REIS, L. L.

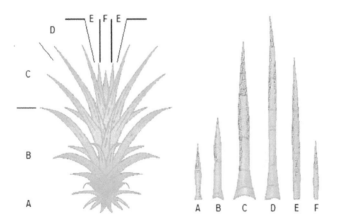

Figure 7. Morphological representation of pineapple tree leaves. Leaf A, the eldest, to F, the youngest. (Adapted from [60]).

As an evaluation parameter for the nutritional state, there are nutrient levels in leaves compared to optimal values, such as in sufficiency ranges or critical levels, presented in tables. Thus, when a nutrient concentration is different from the values presented in those tables, it is suggested that it will limit the plant growth, or productivity and even quality of the fruit.

According to the literature, there are indications of contents of nutrients considered appropriate for the pineapple tree (Table 9). It is observed that there are variations among the nutrients compared to the whole leaf, chlorophylled and achlorophylled portion. This fact shows the importance of standards when collecting leaf samples.

Nutrients	Authors							
	A	B	C	D*	E**	F***	G	H
					g kg⁻¹			
N	10.3	8.8	16.3	13.0-15.0	6.6-9.7	10.9	15.0-17.0	15.0-25.0
P	1.4	1.5	2.1	1.0-1.4	0.3-13.8	2.0	0.8-1.2	1.4-3.5
K	25.0	22.0	20.0	20.0-24.0	3.2-13.8	24.0	22.0-30.0	43.0-65.0
Ca	3.4	3.2	3.9	4.3-7.6	0.9-2.3	6.5	8.0-12.0	2.2-4.0
Mg	3.5	3.1	2.4	2.1-3.6	0.5-1.3	2.2	3.0-4.0	4.1-5.7
S	0.6	0.7	1.3	1.4-1.8	0.4-1.2	1.6	-	-
					mg kg⁻¹			
Fe	73.0	65.0	77.0	-	-	118.0	100.0-200.0	80.0-150.0
Mn	149.0	132.0	67.4	-	-	127.0	50.0-200.0	150.0-400.0
Zn	13.6	14.0	14.3	-	-	12.5	5.0-15.0	15.0-70.0
Cu	-	-	4.5	-	-	4.5	5.0-10.0	10.0-50.0
B	-	-	26.0	18.0-30.0	5.5-8.5	22.0	20.0-40.0	-
Productivity (fruits ha⁻¹ x 1000)								
	40	20	-	-	-	-	-	-
Part of the plant analyzed								
Whole leaf'							Chlorophylled Portion	Achlorophylled Portion

Source: Adapted from [61]; A and B -[62]; C – [63,64]; D*-[65,66], Cultivation with full fertilization; E**- [65,66], Cultivation with nutrient deficiency; F*** -[67], Contents of nutrients in foliar dry matter from pineapple tree seedling at nine months after seed-plotting of sections from the stem; G-[68]; H- [69].

Table 9. Nutrient contents and ranges observed in foliar dry matter from the pineapple tree in different trials.

DRIS[(1] is an alternative technique to evaluate the nutritional state. The critical levels of N, P, K, Ca and Mg were estimated by [70] from the DRIS rules for the "Smooth Cayenne" pineapple tree, in the Bauru – SP region: N (12.0 +/- 0.3[(2)]), P (0.92+/- 0.02), K (21.4+/- 0.6), Ca (4.0+/- 0.1), Mg (2.8 +/- 0.1), where (2) is the confidence interval (95% CI) for foliar critical levels estimated by means of a multiple regression between the DRIS indexes and the levels of macronutrients in the leaves.

---

1 Integrated System of Diagnosis and Recommendations

4.5. Visual diagnosis

The information on visual symptoms of nutritional deficiencies of the pineapple tree, described hereafter, are reported according to [71].

Potassium deficiency (Figure 8) is characterized by green to dark green foliage, more pronounced with nitrogenized fertilization. The leaves show small yellow dots that grow, multiply and may concentrate on the limb margins. Dryness on the apical extremity also occurs. The plant presents erect port and slightly resistant peduncle. The fruit is small, with low acidity and no aroma.

Figure 8. Symptoms of potassium deficiency in pineapple tree leaves. Photo: REINHARDT, D.H.

Potassium deficiency occurs frequently, except in plantings installed in soils rich in this nutrient. It is favored by unbalanced nitrogen rich fertilization, by strong solar radiation, by intense lixiviation and by soils with increased pH and rich in Ca and Mg. According to [72] potassium fertilization intensifies the color of the skin of ripe fruits, changes the color of the pulp from yellow-straw to golden yellow, increases the content of total soluble solids and acidity and improves the organoleptic characteristics of the fruits, providing a better commercial value.

In [47], fruits with lower levels of sugar, less acids, slightly colored, weaker aroma and little resistant peduncle were observed under potassium deficiency, turning those fruits more susceptible to tipping and sun burning. According to [73], it was described that the visual symptoms of K deficiency are characterized by presenting the apex of the older leaves browned and necrotic. Fruits deficient in K presented a pulp with interior darkening.

As general information, it can be stated that the pineapple tree fertilization should be performed in the vegetative phase of the plant cycle, period in which there is a more efficient use of the nutrients applied. Anyway, caution should be exercised regarding the decision making on applying fertilizers in the reproductive phase of the plant cycle, considering the likelihood of increasing the production costs.

4.6. Effect of potassium fertilizer on pineapple

Potassium is the nutrient required in the largest amount by the pineapple tree and its lack represents not only the decrease in plant growth and production, but also affects the quality of the fruits. Facing the importance of this nutrient for the culture, a research was performed where the main focus was to exploit the effects of potassium fertilization in aspects concerning the production and quality of the fruits.

4.7. Methodology

The experiment was conducted in the period from April 2007 to November 2008, in the sector of agricultural production in the State University of Mato Grosso do Sul, located in Cassilândia, MS, with approximately 471m altitude, 19° 05′ S latitude and 51° 56′ W longitude. The climate of the region, according to the classification of [74] is considered rainy tropical (Aw), with a rainy summer and dry winter. Considering the tropical climate in the city of Cassilândia which holds minimal temperatures of 11.19 °C to 22.66 °C and maximum temperatures of 28.35 °C to 36.16 °C with annual precipitation of 2000 mm.

The monthly variations of temperature and precipitation occurring during the conduction of the experiment are represented in Figure 9.

For its execution, the experiment was installed in a medium texture soil, with chemical composition during the implantation period of implantation and conduction of the experiment as presented in Table 10.

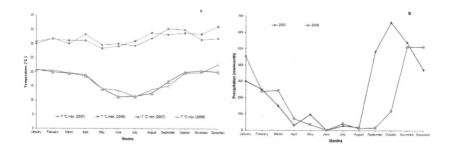

Figure 9. Maximum and minimum temperatures (A) and monthly precipitation (B), during the period of conduction of the experiment. Data provided by Agrometeorological Station from INPE/CPTEC (Instituto Nacional de Pesquisas Espaciais/Centro de Previsão de tempo e Estudos Climáticos).UEMS. Cassilândia, MS, Brazil. 2009.

pH	MO	P	K	Ca	Mg
CaCl$_2$	g dm^{-3}	mg dm^{-3}	-----------------mmol$_c$ dm^{-3}-------------------		
5.0	9.0	5.0	1.0	9.0	3.0
S	H+Al	Al	SB	CTC	V
(mg dm^{-3})	----------------------------mmol$_c$ dm^{-3}------------------------				%
4.0	20.0	3.0	12.6	34.8	36.0
Cu	Fe	Mn	Zn	B	
----------------------------DTPA----------------------------				Warm water	
0.4	21.0	46.0	0.8	0.2	

Source: Instituto Brasileiro de Análises – (IBRA AgriSciences). Campinas, SP, Brazil – 2009.

Table 10. Chemical analysis of the soil in layer from 0 to 20cm. Performed by the Instituto Brasileiro de Análises – (IBRA AgriSciences). UEMS. Cassilândia, MS, Brazil. 2009.

For the installation of the experiment, seedlings of the "Pearl" pineapple tree, from the own University matrix mill located in the production sector. In soil preparation, liming was performed to increase the saturation by bases to 50% and the magnesium content to a minimum of 5 mmol$_c$ dm^{-3}. The experiment was installed in April 2007, with 0.80 X 0.30 m spacing with a density of 41,666.66 plants ha^{-1}, conducted in single lines. The fertilization in the planting furrow with 140 kg ha^{-1} P$_2$O$_5$ (320 kg ha^{-1} triple superphosphate) and top-dressing fertilization performed six times during the performance of the experiment with 222.22 Kg ha^{-1} urea (100 kg ha^{-1} N), in each fertilization, according to soil analysis (Table 11). For the treatments constitution fertilization, the following doses of potassium (K$_2$O) were used: 0, 200, 400, 600, 800 kg ha^{-1}, applied in 4 subdivisions, in June and December 2007; March and June 2008.

The control of weeds, plagues and diseases were made during the whole period of the experiment, with herbicide Glyphosate (1 L ha^{-1}), associated to mechanical methods; application of Imidacloprid insecticide 700 WG (30g 100 L H$_2$O) and Tiametoxam 250 WG (300g 100 L H$_2$O); fungicide Tebuconazole 200 EC (1L ha^{-1}), respectively. The phytosanitary control was performed specifically to preclude problems related to diseases that enable the bad development of the culture. Floral induction was performed at 13 months of age, period in which the plant obtained enough size and age to respond to respond to floral differentiation stimulation. The product applied was Ethrel 720 (720 g etephon L^{-1}), at the dose of 1.0 L ha^{-1}, being performed late in the afternoon to improve the efficiency of the product.

The experimental design used was random blocks with four repetitions and 5 treatments, with experimental division composed by 7 plants, 5 central plants composing the useful portion. The treatments employed are in Table 11.

Treatments	K₂O Rates	KCl Rates *
	----------------------------------Kg ha⁻¹-----------------------------	
1	0.0	0.0
2	200.0	333.3
3	400.0	666.6
4	600.0	1000.0
5	800.0	1333.3

*Potassium chloride (60% K₂O), used in the application of treatments.

Table 11. Treatments used in the experiment with potassium fertilization. UEMS. Cassilândia, MS, Brazil. 2009.

4.8. Results

a. Potassium fertilization and productivity

The results obtained evidenced that the productivity of fruits was influenced by potassium fertilization applied in top-dressing. The regression analysis evidenced a quadratic behavior whose estimated value was 409.38 Kg ha⁻¹ K₂O associated to a productivity of 52,507.30 Kg ha⁻¹ (Figure 10). The quadratic response of the pineapple tree to the KCl doses (Figure 10), may be associated to the depressive effect of the chloride ion, mainly at the higher doses, because according to [75], it is a plant sensitive to chloride toxicity.

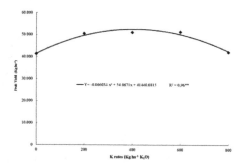

Figure 10. Effect of potassium doses in the productivity of the cultivar Pearl pineapple plant. UEMS. Cassilândia, MS, Brazil. 2009.

According to [42], a positive and significant association was verified within the production of fruits of the "Pearl" pineapple tree. The application of potassium promoted a productivity of 79 t ha⁻¹ of fruits with the estimated dose of 22 g K₂O plant⁻¹.

The potassium content available in the soil and the source used has a promising effect on the way that the pineapple is responsible to potassium fertilization. In an essay performed in the region of Bauru, in Argisoil with $0.7 mmol_c$ dm^{-3} K available, the productivity of the Smooth Cayenne cultivar increased 9.2 t ha^{-1} in response to the application of KCl and 15 t ha^{-1} with the use of potassium sulphate, increases of 18% and 29% related to the control without K application were found, respectively [70]. According to [43, 76], the division of potassium doses in 4 applications during the culture cycle is efficient, regarding the maintenance of fruit quality, and mainly minimizes the losses by lixiviation, increasing the efficiency of the fertilization.

2. Potassium fertilization and fruit quality

In this experiment the quality of the fruits was evaluated from chemical attributes, soluble solids, titratable acidity, *ratio* (SS/AT) and ascorbic acid.

The content of soluble solids estimates the concentration of sugars, which, in most cases, determine the flavor of the fruit. The measurement of soluble solids is used as an indicator of the maturation and quality of the fruits. The fruits destined to *in natura* consumption should have a content of soluble solids higher than 12 ºBrix [77]. In this experiment the influence of potassium on the soluble solids content, which values ranged from 13 to 13.50 ºBrix, was not verified. According to observations performed by [78], the soluble solids contents, found in fruits of Smooth Cayenne fruit tree, which varied from 13.74 to 15.50 ºBrix, were also not directly influenced by K_2O treatments. In a trial performed by [79], the positive effect of potassium fertilization on total soluble solids contents (TSS) was observed, which the application of 490 kg ha^{-1} K_2O increased the TSS in about 6%.

For the characteristic titratable acidity, the data adjusted to a linear ascending equation related to the dose of K_2O. The maximum value of 1.01 mL 100g citric $acid^{-1}$ was found, when associated to the application of 800 Kg ha^{-1} K_2O (Figure 11). Observations made by [80] indicate that the pineapple fruits subjected to low temperatures, both before and after harvesting, increase the incidence and severity of internal darkening of the fruit (*blacheart* or *internal browing*). It has also been observed that the affected fruits are mildly acid and have low levels of ascorbic acid.

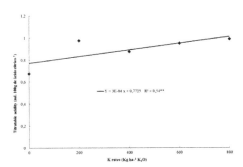

Figure 11. Effect of potassium doses in total titratable acidity of cultivar 'Pearl' pineapple tree fruits. UEMS. Cassilândia, MS, Brazil. 2009.

Regarding the *ratio*, the increment in K_2O doses verified that with the dose increase there was a decrease in the index. The values found were 17.78, with no application of K_2O and 12.78, when the maximum dose was used (800 Kg ha⁻¹). The reduction in the index coupled to potassium dose increments was attributed to a higher increase in acidity related to soluble solids content (Figure 12). This observation corroborates [79], who verified that in Smooth Cayenne pineapple tree, the potassium fertilization acted in two ways in the formation of the *ratio*, as a function of the type of potassium source used. Thus, the increment in titratable acidity was bigger only when KCl was employed as a source of K than the one observed with K_2SO_4 application. Thus, the applications over 400 Kg há⁻¹ K_2O under the form of KCl, implicated in the production of fruits with a *ratio* below 20. When potassium sulphate was employed as K source, it was possible to employ higher K doses, without the *ratio* being lower than 25.

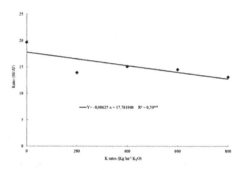

Figure 12. Effect of potassium doses in the *ratio* of cultivar 'Pearl' fruits of pineapple tree. UEMS. Cassilândia, MS, Brazil. 2009.

4.9. Conclusions

With the results obtained, the potassium fertilization coupled with the division of doses met the requirement of the plants, according to its development cycle, mainly at the association of K_2O to the dose of 410 Kg ha⁻¹.

5. Final considerations

For the improvement of the quality of the fruits it is important the use of soil analysis, foliar analysis and visual diagnosis. Those tools are extremely important because they enable the clear evaluation of the availability of potassium and further nutrients available for the plants. Those tools also help in the rational use of fertilizers because besides the producer refraining the waste with excessive fertilization, it does not apply some nutrient that could be limited in production, thus improving the economic results of cultivation with no damages to the environment.

Thereby the efficiency in the use of K is directly related to the direct effects on the amount and quality of fruits, being the handling of the fertilization one of the factors that strongly influences the sustainability in the production of fruit trees. In addition to the choice of K source, strategies should be sought for the handling of solo-plant systems that minimize the loss of this nutrient, taking the example of dose division, minimizing losses by lixiviation.

Author details

Sarita Leonel[1*] and Luis Lessi dos Reis[2*]

*Address all correspondence to: sarinel@fca.unesp.br

*Address all correspondence to: lessireis@fca.unesp.br

1 UNESP. FCA, Department of plant production, Botucatu, SP, Brazil

2 UNESP (São Paulo State University), Botucatu-SP, Brazil

References

[1] Leonel, S. A figueira. Revista Brasileira de Fruticultura, Jaboticabal/SP, v. 30, set. 2008.

[2] Penteado, S. R. Fruticultura de clima temperado em São Paulo. Campinas: Fundação Cargil, 1986. p.115-129.

[3] Nogueira, D.J.P. Nutrição de fruteiras. Informe Agropecuário, Belo Horizonte, v.11. n.125, p.12-31, 1985.

[4] Campo-Dall'orto, F.A. et al. Frutas de clima temperado: II. Figo, maçã, marmelo, pêra e pêssego em pomar compacto. In: RAIJ, B. van. et al. Recomendações de adubação e calagem para o Estado de São Paulo. 2 ed. Campinas: Instituto Agronômico, Fundação, Instituto Agronômico de Campinas, 1996. p.139-140.

[5] Fernandes, F.M.; Buzetti, S. Fertilidade do solo e nutrição da figueira. In: Simpósio Brasileiro Sobre A Cultura Da Figueira, 1999, Ilha Solteira. Anais... Ilha Solteira: FU-NEP, 1999. p.69-85.

[6] Hiroce, R. et al. Composição mineral e exportação de nutrientes pelas colheitas de frutos subtropicais e temperadas. In: Congresso Brasileiro De Fruticultura, 5., 1979, Pelotas. Anais...Pelotas: Sociedade Brasileira de Fruticultura, 1979. p.179-189.

[7] Hernandez, F.B.T. et al. Efeitos de lâminas de irrigação e níveis de nitrogênio sobre os principais parâmetros produtivos da cultura do figo (Fícus carica L.). In: CON-

GRESSO BRASILEIRO DE ENGENHARIA AGRÍCOLA, 21., 1992, Santa Maria. Anais...Santa Maria: Sociedade Brasileira de Engenharia Agrícola, 1992. v.2B, p. 875-885.

[8]   Brizola, R.M.; Leonel, S.; Tecchio, M.A.; Hora, R.C. da. Teores de macronutrientes em pecíolos e folhas de figueira (Ficus carica, L.) em função da adubação potássica. Ciência e Agrotecnologia, Lavras, v. 29, n. 3, p. 610-616, 2005.

[9]   Tubelis, A.; Nascimento, E.J.L.; Foloni, L.L. Meteorologia e climatologia. Botucatu. Faculdade de Ciências Médicas e Biológicas, 1972. v.3, p.344-362 Mimeografado.

[10] Curi, P.R. Relações entre evaporação média pelo tanque IA-58 e evapotranspiração calculada pelas equações de Thornthwaite e Camargo, para o município de Botucatu. 1972. 88 f. Tese (Doutorado) – Faculdade de Ciências Médicas e Biológicas de Botucatu, Universidade Estadual Paulista, Botucatu, 1972.

[11] United States Soil Conservation. Soil Taxonomy: A Basic System of Soil Classification for Making and Interpreting Soil Surveys. 1988. 754 p.

[12] Raij, B. V.; Quaggio, J. A. Métodos de análise de solo para fins de fertilidade. Campinas: Instituto Agronômico de Campinas, 1983. (Boletim Técnico, 81).

[13] Epstein, R.P. Nutrição mineral das plantas: princípios e perspectivas. São Paulo: Livros Técnicos e Científicos Editora da Universidade de São Paulo, 1975. 314 p.

[14] Malavolta, E. Nutrição mineral. In: Ferri, M.G. Fisiologia vegetal. 2. ed. São Paulo: Editora da Universidade de São Paulo, 1985. cap.2/3, p.77-116.

[15] Mengel, K.; Kirkby, E.A. Principles of plant nutrition. 4.ed. Bern: International Potash Institute, 1987. 655 p.

[16] Haag, H. P. et al. Distúrbios nutricionais em figueira (Ficus carica L.) cultivada em solução nutritiva. O solo, Piracicaba, v.71, n.1, p.31-34, 1979.

[17] Malavolta, E.; Vitti, G. C.; Oliveira, S. A. Avaliação do estado nutricional das plantas: princípios e aplicações. 2. ed. Piracicaba: Associação Brasileira de Potassa e do Fósforo, 1997. 319 p.

[18] Quaggio, J.A.; Raij, B. van.; Piza JR.C de T. Frutíferas. In: RAIJ, B. van. et al. Recomendações de adubação e calagem para o Estado de São Paulo. 2. ed. Campinas: Instituto Agronômico, Fundação, IAC, 1996. p.121-153.

[19] Pereira, J.R. et al. Nutrição e adubação da videira. In: LEÃO, P.C.S.; SOARES, J.M. A vitivinicultura no Semi-Árido brasileiro. Petrolina: Embrapa Semi-Árido, 2000. cap.9, p.213-258.

[20] Fachinello, J.C.; Manica, I.; Machado, A.A. Respostas da figueira (Fícus carica L.) cv. São Pedro a dois níveis de adubação com nitrogênio, fósforo e potássio. In: Congresso Brasileiro De Fruticultura, 5., 1979, Pelotas. Anais... Pelotas: Sociedade Brasileira de Horticultura, 1979. v.3, p. 889-895.

[21] Petri, J.C. Macieira. Revista Brasileira de Fruticultura, Jaboticabal, v. 30, n. 4, 2008.

[22] Ernani, P.R.; Dias, J. Flore, J.A. Annual aditions of potassium to the soil increased apple yield in Brazil. Communication in Soil Science and Plant Analysis, v. 33, p. 1291-1304, 2002.

[23] Petri, J.L. Formacao de flores, polinizacao e fertilizacao. In: EPAGRI. Manual da cultura da macieira. Florianópolis: Epagri, 2002. p. 229-259.

[24] Taiz, Z.; Zeiger, E. Fisiologia vegetal. Porto Alegre: Artmed, 2004. 719 p.

[25] Neilsen, G.H.; Parchomchuk, P.; Meheriuk, M.; Neilsen, D. Development and correction of K-deficiency in drip irrigated apple. Hortscience, v. 33, p. 258-261, 1998.

[26] Daugaard, H.; Grauslund, J. Fruit colour and correlations with orchard factors and post-harvest characteristics in apple cv Mutsu. Journal of Horticultural Sciences & Biotechnology, v. 74, p. 283-287, 2000.

[27] Neilsen, G.H.; Parchomchuk, P.; Neilsen, D.; Zebarth, B.J. Drip-fertirrigation of apples trees affects root distribution and the development of K deficiency. Canadian Journal of Soil Science, V. v. 80, P. 353-361, 2000.

[28] Hunsche, M.; Brackmann, A.; Ernani, P.R. Efeito da adubacao potassica na qualidade pos-colheita de macas 'Fugi'. Revista Agropecuária Brasileira, v. 38, p. 489-496, 2003.

[29] Nava, G.; Dechen, A.R.; Natchigal, G.R. Nitrogen and potassium fertilization affect apple fruit quality in Souther Brazil. Communication in Soil Science and Plant Analysis, v. 39, p. 96-107, 2008.

[30] Argenta, L.C. Fisiologia pos-colheita: maturação, colheita e armazenagem dos frutos. In: EPAGRI. Manual da cultura da macieira: Florianópolis: Epagri, 2002. p. 691-732.

[31] Comissao De Quimica E Fertilidae Do Solo RS/SC. Manual de adubação e calagem para os estados do Rio Grande do Sul e de Santa Catarina. 10 ed. Porto Alegre: CQFS-RS/SC, 2004. 400 p.

[32] Nava, G., Dechen, A.R. Scientia Agrícola, Piracicaba, v. 66, n. 3, p. 377-385, May/June 2009.

[33] PY, C.; Lacoeuilhe, J.J.; Teisson, C.L. L'Ananas sa culture, ses produits. Paris: G. M. Maisoneuve et Larose, 1984, 562p.

[34] Reindhardt, D. H.; Souza, J. da. S. Pineapple industry and research in Brazil. Acta Horticulturae, Wageneingen, n. 529, p. 57-71, 2000.

[35] Bengozi, F. J. Procedência, sazonalidade e qualidade físico-química do abacaxi comercializado na CEAGESP – São Paulo. 2006. 151 f. Dissertação (Mestrado em Agronomia – Horticultura) – Universidade Estadual Paulista "Júlio de Mesquita Filho", Botucatu, 2006.

[36] Cunha, G. A. P.; Cabral, J. R. S. Taxonomia, espécies, cultivares e morfologia. In: Cunha, G. A. P.; Cabral, J. R. S.; Souza, L. F. S. O abacaxizeiro: cultivo, agroindústria e

economia. Brasília: Embrapa Comunicação para Transferência de Tecnologia, 1999. p. 17-28.

[37] Manica, I. Fruticultura tropical:5. Abacaxi. Porto Alegre: Cinco Continentes, 1999. 501p.

[38] Hepton, A. Cultural System. In: Bartholomew, D.P.; Paul, R.E., Rohrbach, K.G. The Pinneaple- Botany, Production and Uses. Honolulu: CABI Publishing, 2003. p-109-142.

[39] FAO, Faostat – FAO statistical data bases. Roma: World Agricultural Information Centre, 2011. Disponível em: <http://apps.fao.org>. Acesso em: 26 nov. 2011.

[40] Ibge, Sidra. Disponível em: <www.sidra.ibge.gov.br/bda> Acesso em 26 nov. 2011

[41] Silva, A.P.; Souza, A.P.; Alvarez V., V.H.; Dantas, J.P.; Celestino, A.P.Q.; Oliveira, F.P. Estudo das relações entre Ca, Mg, K e CTC em solos da região abacaxicultora do Estado da Paraíba. In: Fertbio 2004 – Reuniao Brasileira De Fertilidade Do Solo E Nutricao De Plantas, 26, Lages-SC, 2004.

[42] Veloso, C.A.C.; Oeiras, A.H.L.; Carvalho, E.J.M.; Souza, F.R.S. Resposta do abacaxizeiro a adição de nitrogênio, potássio e calcário em Latossolo Amarelo do Nordeste Paraense. Revista Brasileira de Fruticultura, Jaboticabal, n. 23, p. 396-402, 2001.

[43] Teixeira, L. A. J.; Spironello, A.; Furlani, P. R.; Sigrist, J. M. M. Parcelamento da adubação NPK em abacaxizeiro. Revista Brasileira de Fruticultura, Jaboticabal, v.24, p. 219 - 224, 2002.

[44] Spironello, A.; Quaggio, J.V.A.; Teixeira, L. A. J.; Furlani, P. R.; Sigrist, J. M. M.. Pineapple yield and fruit quality effected by NPK fertilization in a tropical soil. Revista Brasileira de Fruticultura 26, 155–159, 2004.

[45] Soares, A.G.; Trugo, L.C.; Botrel, N.; Souza, L.F.S. Reduction of internal browning of pineapple fruit (Ananas comosus L.) by preharvest soil application of 139 potassium. Postharvest Biology and Technology, Pullman, v.35, p. 201-207, 2005.

[46] Paula, M.B. DE; Mesquita, H.A. DE; Nogueira, F.D. Nutrição e adubação do abacaxizeiro. Informe agropecuário, v.19, n.195, 1998. p.33-39.

[47] Malézieux, E., Bartholomew, D.P. Plant Nutrition. In: Bartholomew, D.P.; Paul, R.E., Rohrbach, K.G. (eds.) The Pinneaple- Botany, Production and Uses. Honolulu: CABI Publishing, 2003. p.143-165.

[48] PY, C.; Lacoeuilhe, J.J.; Teisson, C. The pineapple, cultivation and uses. Paris: G.P. Maisonneuve & Larose, 1987. 568p.

[49] Souza, L. F. S. Correção de acidez e adubação. In: Cunha, G. A. P.; Cabral, J. R. S.; Souza, L. F. S. O abacaxizeiro: cultivo, agroindústria e economia. Brasília: Embrapa Comunicação para Transferência de Tecnologia, 1999. p.169-202.

[50] Malavolta, E. Nutrição mineral e adubação do abacaxizeiro. Anais do Simpósio Brasileiro de Abacaxizeiro,1ed., Jaboticabal: FCAV, 1982. p.21-153.

[51] Paula. M. B.; Carvalho, J. G.de; Nogeura, F. D.; Silva, C. R. Exigências nutricionais do abacaxizeiro. Informe agropecuário, Belo Horizonte, v. 11, n. 1, p. 27-32. 1985.

[52] Souza, L. F. S. Adubação. In: Reinhardt, D. H.; Souza, L. F. da S.; Cabral, J. R. S. Abacaxi. Produção: Aspectos técnicos. Brasília: Embrapa Comunicação para Transferência Tecnológica, 2000. p. 30-34. (Frutas do Brasil, 7).

[53] Brasil. Ministério da Agricultura, Pecuária e Abastecimento. Instrução Normativa nº 5, de 23 de fevereiro de 2007.

[54] Paula, M.B. DE; Mesquita, H.A. DE; Nogueira, F.D. Nutrição e adubação do abacaxizeiro. Informe agropecuário, v.19, n.195, 1998. p.33-39.

[55] Bezerra, J.E.F.; Lederman, I.E.; Abramof, L.; Cavalcante, U.M.T. Sulfato e cloreto de potássio na produtividade e qualidade do abacaxi cv. Smooth Cayenne. Revista Brasileira de Fruticultura, Cruz das Almas, n.5, p.15-20, 1983.

[56] Bezerra, J.E.F.; Maaze, V.C.; Santos, V.F.; Lederman, I.E. Efeito da adubacao nitrogenada, fosfatada e potassica na producao e qualidade do abacaxi cv. Smooth Cayenne. Revista Brasileira de Fruticultura, Cruz das Almas, n. 3, p.1-15, 1981.

[57] Paula, M.B.; Carvalho, V.D.; Nogueira, F.D.; Souza, L.F.S. Efeito da calagem, potassio e nitrogenio na producao e qualidade do fruto do abacaxizeiro. Revista Brasileira de Fruticultura, Jaboticabal, n. 26, p.1337-1343,1991.

[58] Ernani, P.R.; Almeida, J.A.;Santos, F.C. Potássio. In: Novais, R.F. et al. (Ed.). Fertilidade do solo.Viçosa:SBCS, 2007.p.551-594.

[59] Lacoeuilhe, J.J. La fumure N-K de l'ananas en Côte d'Ivoire. Fruits, v. 33, p.341-348, 1978.

[60] Py, C. La pinã tropical. Barcelona: Blume, 1969. 278p.

[61] Freitas, M.S.M.; Carvalho, A.J.C.; Monnerat, P.H.; Diagnose foliar nas culturas do maracujá e do abacaxi. In. Prado, R.M. (Ed.). Nutrição de plantas: Diagnose foliar em frutíferas. Jaboticabal, SP: Fcav/Capes/Fapesp/CNPq, 2012. p. 227-258

[62] Faria, D. C. DE. Desenvolvimento e produtividade do abacaxizeiro 'Smooth Cayenne' em função de adubação nitrogenada e tipos de mudas no Norte Fluminense. 2008. 67f. Dissertação (Mestrado em Produção Vegetal)-Universidade Estadual do Norte Fluminense Darcy Ribeiro, Campos dos Goytacazes-RJ.

[63] Siebeneichler, S.C.; Monerat, P.H.;Carvalho, A.J.C. DE.; Silva, J.A. DA. Composição mineral da folha em abacaxizeiro: efeito da parte da folha analisada. Revista Brasileira de Fruticultura, v. 24, p. 194-198, 2002.

[64] Siebeneichler, S.C.; Monnerat, P.H.; Silva, J.A. DA. Deficiência de boro na cultura do abacaxi "Pérola". Acta Amazonica, v.38, p.651-656, 2008.

[65] Ramos, M.J.M.; Monnerat, P.H.; Carvalho, A.J.C. DE; Pinto, J. L.A.; Silva, J.A. DA. Nutritional deficiency in 'Imperial' Pineaple in the vegetative grownt phase and leaf nutrient concentration. Acta Horticulturae, v. 702, p. 133-139, 2006.

[66] Ramos, M. J.M.; Monnerat, P.H.; Pinho, L.G. DA R.; Silva, J.A. DA. Deficiência de macronutrientes e de boro em abacaxizeiro 'Imperial': composição mineral. Revista Brasileira de Fruticultura, v.33, p.261-271, 2011.

[67] Coelho, R.I.; Carvalho, A.J. C.; DE; Thiebaut, J.T.L.; Souza, M.F. Teores foliares de nu-trients em mudas do abacaxizeiro Smooth Cayenne em resposta à adubação. Revista de Ciências Agrárias, v. 33, p.173-179, 2010.

[68] Boaretto, A.E.; Chitolina, J.C.; Raij, B. Van; Silva, F.C.;Tedesco, M.J.; Carmo, C.A.F.S. Amostragem, acondicionamento e preparação das amostras de plantas para análise química. In: SILVA, F.C. da. (org) Manual de análises químicas de solos, plantas e fertilizantes. Brasília: Embrapa, 1999. P.49-74.

[69] Reuter, D.J.;Robinson, J. B. Plant analyses – na interpretation manual. Melbourne: Inkata Press, p.218.1988.

[70] Teixeira, L.A.J.; Quaggio, J.A.; Zambrosi, F.C.B. Preliminary Dris norms for "Smooth Cayenne" pineapple and derivation of critical levels of leaf nutrient concentrations. Acta Horticulturae, n.822, p.131-138, 2008.

[71] Matos, A.P. de. (organizador). Abacaxi. Fitossanidade. Brasília: Embrapa Comunica-ção para Transferência de Tecnologia, 2000. 77 p.; il; (Frutas do Brasil; 9)

[72] Unsherwood, N.R. The role of potassium in crop quality. In: Munson, R.D. (Ed.). Po-tassium in agriculture. Madison: ASA/CSSA/SSSA, 1985. p.489-513.

[73] Ramos, M.J.M.; Monnerat, P.H.; Carvalho, A.J.C. DE; Pinto, J.L. DE A.; Silva, J.A. DA. Sintomas visuais de deficiência de macronutrientes e de boro em abacaxizeiro "impe-rial". Revista Brasileira de Fruticultura, v.31, p.252-256, 2009.

[74] Köeppen, W. Roteiro para classificação climática. [S. I. : s. n.], 1970. 6 p. (não publica-do, mimeografado).

[75] Teiwes, G.; Gruneberg, F. Conocimientos y experiências en la fertlización de la piña. Boletín Verde, v.3, p.1-67, 1963.

[76] Giacomelli, E.J.; PY, C.O abacaxi no Brasil. Campinas: Fundação Cargill, 1981. 101p.

[77] Brasil. Ministério da Agricultura, Pecuária e Abastecimento. Instrução Normativa/ SARC nº 001, de 1º de fevereiro de 2002.

[78] Maeda, A. S. Adubação foliar e axilar na produtividade e qualidade de abacaxi. 2005. 43 f. Dissertação (Mestrado em Sistemas de Produção) – Universidade Estadual Pau-lista "Júlio de Mesquita Filho", Ilha Solteira, 2005.

[79] Quaggio, J. A.; Teixeira, L.A.J.; Cantarella, H.; Mellis, E.V.;Sigrist, J.M. Post-harvest behaviour of pineapple affected by sources and rates of potassium. Acta Horticulturae, 822:277-280, 2008.

[80] Teisson, C.; Lacoeuilhe, J.J.;Combres, J.C. Le brunissement interne de l'ananas.V. Recherches des moyens de lutte. Fruits, v.34, p.399-415, 1979.

Chapter 8

Sustainable Management of Soil Potassium – A Crop Rotation Oriented Concept

Witold Grzebisz, Witold Szczepaniak,
Jarosław Potarzycki and Remigiusz Łukowiak

Additional information is available at the end of the chapter

1. Introduction

Modern agriculture is under pressure of two contradictory challenges reflected by the increasing world's population on one hand, and the magnitude of food production, on the other hand. In the period ranging from 1960 to 2010, the population doubled from 3 to more than 6 billions, while the production of cereals tripled, a success which expressed by a significant yield increase per ha (from 1.09 t ha^{-1} in 1960 to 3.0 t ha^{-1} in 2010) [1]. The major reason of such yield increase was a marked progress in plant breeding, resulting in generations of new, high-yielding varieties [2]. This process run in parallel with the increase in fertilizers, pesticides production and consumption, hence enabling to cover nutritional needs and supporting the health of high-yielding crops. The intensive production gain, based on enormous consumption of non-renewable resources, especially fuel and simultaneously nutrients such as nitrogen and phosphorus was, however, concomitant with their low use efficiency. This type of agriculture intensification created, in many regions of the world a threat to environment, at both local and global-scale. There are numerous examples stressing the negative impact of intensive agriculture on environment. Agricultural practices are responsible for the majority of ammonia and to a great part for nitrogen oxide's emission to the atmosphere. Pollution of ground-water by nitrates and phosphates originating from both arable soils and surface waters was recognized the earliest. All these negative effects were the reason for the increased activity of local societies in the 70 and 80-ies of the XX century, resulting in the development of legal instruments protecting the environment, for instance the Nitrate Directive [3, 4, 5].

The complexity of agricultural impact on human life and the increasing awareness of environmental threats was the boosting argument for elaborating a concept of sustainable agriculture growth [6]. The core of this concept relies on an assuming that agricultural systems should be managed in a way covering current needs of present human being's population without negative impact on its performance in the future. The change of classical technologies to fulfill both goals cannot be, however, achieved discrediting existing production methods. The analysis of two food supply scenarios, developed in the 90-ies, i.e., Yield Oriented Agriculture (YOA) and Environment Oriented Agriculture (EOA), implicitly shows that the second scenario guaranties the only moderate diet in 2040. By following this food production strategy, food shortage is expected in some regions of the world [7]. Therefore, the main challenges of modern agriculture are related to the improvement of classical production technologies. The future development scenarios cannot follow the concept of "sustainable intensification." This strategy, developed for low-input agriculture, assumes the substitution of external inputs by naturally available resources, both physical and human [8]. Therefore, the key challenge of agriculture is to increase resources use efficiency in all systems, independently on their current intensity. The general strategies of technological changes should include: i) improvement of water unit productivity, ii) increasing size of soil natural pools, i.e., resources affecting its fertility (organic carbon, macro- and micro-nutrients), iii) reorientation of plant crop management of a single crop to the crop rotation iv) adopting no-till farming and conservation agriculture [2, 5, 9, 10, 11].

The efficient allocation of production means for improving yields and securing the environment, requires a deep insight into processes responsible for crop's productivity. It is well recognized, that crop plant development during the growth season is controlled by numerous factors, both depended and independent on farmer's activity. All these factors have been arranged in a manner taking into account the degree of their impact on plant growth and productivity [12]. Four hierarchical levels of production factors and respective yield levels may be distinguished: i) potential, ii) water limited, iii) nutrient limited, iv) actual [13]. The first level of crop plant productivity is defined by climatic factors such as solar radiation, fixed by geographical location of the field. The potential productivity of the presently cultivated variety is defined by the capacity of its canopy to intercept solar radiation. This yield category is achievable, provided ample supply of water and nutrients during all stages of yield development [14]. For example, in Europe, the average yield potential of wheat was evaluated for the period 1996-2005 at the level of 10.4 t ha^{-1}, however, ranged from 6.9 in Bulgaria to 12.7 t ha^{-1} in Ireland [15], depending on climatic factors.

Lal [10] has made an important remark concerning the exploitation of the yield potential of modern species. He pointed out that …"improved germplasm cannot extract water/ nutrients from degraded/depleted soils".… Water supply to plants during the vegetative season is considered as the key limiting factor, defining the maximum achievable yield under physical conditions of the currently cropped field. In other words, this factor determines the site-specific, i.e., locally realizable yield. Any shortage of water supply throughout the growth season, especially during critical stages of yield formation is the primary reason of yield losses. In Europe, the water limited yield (WLY) is fixed at levels, showing declining trends in the directions extending from the West to the East and the South of the continent. For Ireland, it has been calculated at the level of 8.5, Germany and Poland – 6.5, Bulgaria, Romania – 4.5 t ha^{-1} [16]. However, WLYs show significant differences in comparison to the respective yield potential, which is as follows: 4.2, 3.4, 3.6, 4.6, 2.4 t ha^{-1}, respectively. This virtually un-harvested portion of the potential yield has been termed as the yield gap.

The third level of crop productivity depends on the supply of nutrients. However, production effects of applied nutrients are different, depending on a particular yield forming characteristic. Therefore, they can be classified to one of two main groups. The first one comprises only one nutrient, i.e., nitrogen. Its superiority over others is due to the decisive impact on primary plant physiological processes. The most important are those responsive for dry matter production, and its subsequent partitioning among organs during the whole life cycle of a plant [17]. Therefore, water and nitrogen are considered as the key limiting factors in the realization of the crop yield potential.. The effect of both factors on in-season crop performance depends on the supply of other nutrients. All of them are essential for the adequate plant growth, but are considered only as "secondary" in terms of their impact on yield performance and yielding potential exploration. Therefore, the Nutrient Limited Yield Gap (NLYG) can be related to the degree of both water- and nitrogen-use efficiency, i.e, WUE and NUE, respectively. The first one creates a milieu for nitrogen uptake and its further internal utilization. Thereby, the yield gap, due to inadequate uptake of nitrogen can be overcome provided the balanced supply of other nutrients. The question, remains how to match a crop demand for nitrogen and other nutrients in time and space?

The main assumption of efficient nitrogen use is to apply nitrogen fertilizer in accordance to crop plant demands, which are variable during consecutive stages of growth. Farmers are aware of nitrogen and other nutrient's importance for increasing yield of growing crops as a prerequisite of high yield. However, they frequently make savings of their use, in turn decreasing nitrogen production efficiency. The key attribute of nitrogen-oriented crop production is its relatively low recovery from applied fertilizers, in turn negatively impacting the environment [4, 9]. In addition, the unbalanced nitrogen use leads to the depletion of natural resources of other nutrients required by crop plants. This situation, as shown in Fig. 1, is typical for countries of the second and third group. In many Central-East European countries, yields of wheat decreased significantly in the 90-ies. The declining soil fertility is the main cause of the considerable year-to-year variability of harvested yields. The first step in reorientation of current agriculture production into a sustainable way should, therefore, rely on the improvement of phosphorus and potassium management. The best example of this trend is China, which doubled during the last 20 years potassium consumption, resulting in a linear yield increase. The main goals of crop plants fertilization with potassium are to: i) reduce year-to-year variability of harvested yields and ii) increase water- and N-use efficiency.

Potassium is one of the most important nutrients required by crop plants. In plants, its accumulation rate during early stages of growth precedes nitrogen accumulation. Therefore, its supply to plants seems to be decisive for nitrogen utilization, in turn significantly affecting plants growth rate and the degree of yield potential realization. The current status of potassium management in world's agriculture, as presented in Table 1 and Fig. 1, has been evaluated on the basis of potassium fertilizers consumption. Wheat has been considered as an example for assessing, the importance of this nutrient for food production. The consumption of potassium fertilizers in the period from 1986 to 2009 underwent significant changes on the world agricultural scene. The top wheat producers, mainly European countries and USA, showed a sharp declining trend in potassium usage in 2006-2009 as compared to 1986-1990. In 2005-2009, the group of main potassium consumers, compared to the period 1986-1990 decreased its use to 47%. Unexpectedly, in all these countries, any significant negative effect on wheat yields as induced by the decline in the consumption of potassium was noted. For example, grain yield increase averaged over all studied countries, amounted to 0.9 t ha^{-1}, ranging from -0.3 for Denmark to +2.17 t ha^{-1} for Belgium (Table 1). Therefore, it can be concluded, that the recommended rates of potassium fertilizer did not fit real wheat requirements, both in time and space.

Potassium consumption pattern for Central and East European countries is much more complicated. It usage showed the same declining trend as in the previous group. However, in the second period, the average K consumption dropped down to 14.3% of its primary level. The mean change of wheat yield showed increase only for the Russian Federation and stagnation for the Czech Republic. In other countries, a temporary yield gap (TYG), i.e., induced by the decrease in fertilizer's consumption ranged from -5% for Serbia to -23% for Bulgaria. The relative change (ΔY) of wheat yield as presented below, followed the degree of potassium consumption change (ΔK):

$$\Delta Y = 1.47\Delta K + 114.2; R^2 = 0.66, n = 7 \text{ and } P \leq 0.01 \tag{1}$$

The third group consists of low potassium fertilizer consumers (based on data for the 1986-1990 period). Most of them showed an extremely huge K consumption increase in the period extending from 1990 to 2009. This high progress resulted in the net yield of wheat gain, as presented below:

$$\Delta Y = 0.13\Delta K + 4.83; R^2 = 0.61, n = 7 \text{ and } P \leq 0.01 \tag{2}$$

Statistical characteristics	HP[2]		I[3]		P[4]	
	1986-90	2005-2009	1986-90	2005-2009	1986-90	2005-2009
Average	51.4	23.6	53.9	7.75	4.85	14.5
Standard deviation	30.2	14.7	24.7	4.46	4.46	15.4
Coefficient of variation, %	58.8	62.3	45.8	57.5	91.9	106.1

[1]source: FAOSTAT, IFADATA, available online 2012-08-07;

[2]group HP (high productive countries): Austria, Belgium, Denmark, France, Germany, Italy, The Netherlands, Spain, United Kingdom, United States of America;

[3]group I (intermediate): Bulgaria, Czech Republic, Hungary, Romania, Russian Federation, Serbia, Slovak Republic, Ukraine;

[4]group P (progressive): Argentina, Australia, China, Egypt, India, Mexico, Turkey.

Table 1. Statistical overview of potassium consumption by wheat producers in two distinct periods[1], kg K$_2$O ha^{-1}

These three examples, presenting potassium management by main wheat producers, implicitly indicate that there is a significant gap between official K recommend rates and real needs of wheat for fertilizer potassium. It is necessary to agree with opinions expressed frequently by farmers, about the inappropriateness of current nutrient recommendations and especially regarding the transfer of scientific knowledge to agriculture practice. Consequently, each method of N management requires, firstly, a simple and secondly, a reliable method of other nutrient's recommendations in terms of the amount and of time, as the guarantee of available N efficient use.

This conceptual review assumes that sustainable potassium management on the field should focus farmer's activity on increasing both: i) the amount of available K pool and ii) access of crops grown in a given cropping sequence to this resources. The primary objective of this paper is to present and explain the scientific background of potassium impact on crop plant's growth and productivity, taking into account their different sensitivity to K supply, both in the required quantity and time. The key objective is to stress the importance of the crop rotation-oriented strategy of potassium management, considered as the low input method. It focusses on covering K requirements of the most sensitive crop during its critical stages of yield formation. It is also supposed, that K soil sufficiency can be partly achieved by recycling of organic K sources, taking into account the crop rotation course.

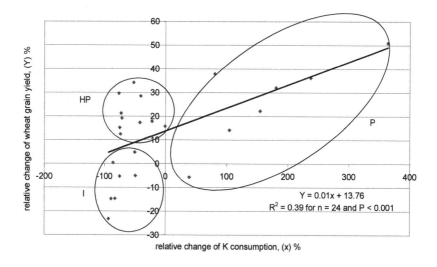

Legend: HP, I, P – groups of countries

Figure 1. Effect of relative change of potassium fertilizer consumption on relative wheat yield change (1986-1990 = 100%).

2. Potassium impact on crop plant productivity – Physiological backgrounds

Potassium (K) is one of the 16 elements needed for plant growth. It is essential in nearly all processes required to sustain the adequate plant growth and its reproduction. Potassium plays a basic role in series of fundamental metabolic and physiological processes in the plant. Plants under potassium deficiency reduce carbon dioxide assimilation and ATP production. Carbon fixation and assimilates transportation to other organs requires potassium. A sufficient supply of potassium is therefore, a background of efficient solar-energy use [19, 20].

A high-yielding crop takes up large quantities of potassium to cover its requirements during the whole vegetation. The highest accumulation of potassium is generally attributed to root crops such as sugar beet and potato. In fact, the first one yielding at the potential level, i.e., 80 t ha^{-1}, accumulates more than 400 kg K ha^{-1} [21]. Cereals are considered as low K consumers. Winter plants yielding at the level of 10 t ha^{-1}, can accumulate at harvest 190 kg K ha^{-1} [22]. It is necessary to stress that the total amount of K in the crop at harvest is by $^1/_{10}$ to $^1/_3$ lower than its maximum during the growth season. This difference should be taken into account when calculating the K recommended rate.

Potassium management by a crop requires an insight into some canopy characteristics: i) quantity of accumulated nutrient, ii) absolute/relative uptake rate, iii) dynamics of nitrogen uptake. The first step in understanding K impact on a crop productivity is to define its sensitivity to K supply. The general trend of K accumulation during the life cycle of a crop can be described by the sigmoid-like curve (Fig. 2). The same patterns, as shown in Fig. 2, have been found for winter wheat [22, 23] and for oil-seed rape [24]. The well-defined maximum describes the date of the highest amount of K fixed by the canopy followed by a subsequent decrease during maturation. The second important information drawn from Fig. 2 refers to K_{max}. As a rule, K accumulation precedes the absolute rate of both dry matter and N accumulation. Based on the pattern of N and K in-season accumulation, it can be formulated a hypothesis, that K accumulated in *excess* during the vegetative part of the seed crop growth builds-up a nutritional buffer, supporting effective N use during the grain filling period. The best examples are cereals, for which the crucial stage of dry matter production occurs at the end of booting and during heading. This period is decisive for establishing both the number of ears and grains per ear. At these stages, cereals reach the top of K accumulation, which is conclusive for high-yield [22, 23, 25].

The dynamics of potassium uptake by a crop can be described using indices such as the absolute/relative rate of K accumulation (A/R-RKA). The first one is shown in Fig. 3 for sugar beet. This crop can keep the uptake rate at the level of 10 and 9 kg K ha^{-1} • d^{-1} for 7 and 17 consecutive days, respectively. Dynamics of K uptake coincides with the absolute rate of the root system extension, reaching top values at the period of maximum dry matter accumulation, both in leaves and roots [26]. In oil-seed rape dominates the same pattern of K and N uptake. The uptake rate of K during the period from the rosette stage up to flowering ranges from 3 to 7 kg K ha^{-1} • d^{-1}, reaching the maximum at booting [24, 27]. The same potassium

accumulation course was found for winter wheat [23]. In this particular case, the highest accumulation rate of K was lower than that observed for leafy crops, achieving 4.4. kg K ha^{-1} •
d^{-1} from the beginning of stem elongation up to heading.

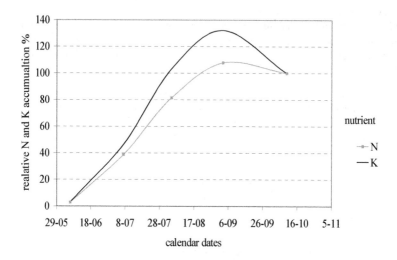

Figure 2. General pattern of K and N accumulation in high-yielding crops, a case of sugar beet; Source [21]

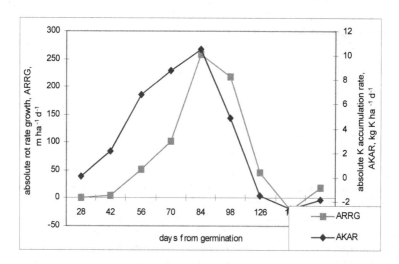

Figure 3. Dynamics of K accumulation by sugar beet on the background of root growth; Adapted from [21, 28]

Based on these sets of data it can be formulated a hypothesis that an efficient supply of potassium to a crop is a prerequisite of achieving the highest rate of canopy growth. The importance of potassium management for dry matter accumulation by a maize canopy is presented in Fig 4. The analysis of the course of crop growth rate (CGR) can be used to discriminate the critical stage of a particular crop response to the supply of potassium. In maize, for example, the elevated rate of dry matter accumulation takes place from tasselling (BBCH 51) and extends up to the blister stage (BBCH 71). This crop shows a very high plasticity to K management. The highest CGR was an attribute of both groups of plants grown on i) a fertile K soil, irrespectively on current K supply, and ii) medium K fertile soil but freshly fertilized with K.

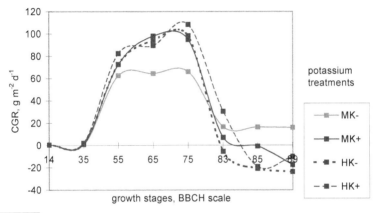

Legend: M, H – levels of soil fertility, medium and high, respectively; K-, K+ - freshly applied potassium; CGR – crop growth rate.

Figure 4. Dynamics of dry matter accumulation by maize on the background of potassium fertilizing; Source [29]

Plants take up potassium as the K^+ ions. Its availability and the plant uptake rate is affected by several soil and plant factors: i) K concentration in the soil solution, ii) size of the soil cation exchange complex, iii) soil properties such as: moisture, soil aeration and oxygen level, temperature, iv) plant crops internal requirements, v) rooting depth [30, 31]. The first two factors are decisive for potassium resources. However, K utility by a particular plant from different soil reservoirs depends on the internal plant requirement, which is defined by the rate of dry matter accumulation, expressed as the biomass ingrowths rate. In fact, it is the basic factor depending on the supply of water and nitrogen. Nevertheless, the evaluation of factor's hierarchy implicitly shows, that the rate of root growth affects the K uptake rate from soil resources the most [32]. The elucidation of the role of root growth requires a deep insight into mechanisms of potassium uptake by a plant root. Its has been well documented that K^+ ion's transportation from the soil solution to the root is mainly *via* diffusion. The movement of potassium depends on the water regime and plant root system activity [33]. The rate of any ion transportation to the root surface is governed by its effective diffusion rate (D_{eff}), which is nutrient specific. Under constant physical conditions, ions with higher D_{eff} are taken faster, resulting in steeper depletion of its concentration in the bulk soil surrounding the root surface [34]. The occurring processes can be described quantitatively using the following set of equations:

$$d = (2 D_{eff} \times t)^{0.5} \tag{3}$$

$$D_{eff} = D_w \times \theta \times f/b \tag{4}$$

$$r_1 = (4/\pi)^{0.5} \times 1/(Lv_r^{0.5}) \tag{5}$$

$$Dd = (d/0.5r_1) \times 100\% \tag{6}$$

where:

d - the root depletion zone, cm

t - time since initiation of calculation, days

D_{eff} - coefficient of diffusion in soil solution for a particular nutrient, cm^2 s^{-1}

D_w - coefficient of diffusion in water for a particular nutrient, cm^2 s^{-1}
θ - volumetric soil water content, cm^3 cm^{-3}

f - tortuosity factor of soil pores, unitless

b - soil buffering capacity for a particular nutrient, unitless

Lv_r - root length density, cm cm^{-3}

r_1 - the mean distance between neighboring roots, cm

Dd - degree of nutrient utility in the depletion zone, %

The typical values of coefficients of diffusion for two main nutrients, i.e., nitrate nitrogen and potassium, are as follows: 2 10^{-6} cm^2 s for nitrates and 2 10^{-7} s for potassium [35]. However, their values may significantly decrease under conditions of low water content down to 10^{-10} and 10^{-12} for both nutrients, correspondingly. The depletion zone calculated using typical values, and a period of seven days extends from the root surface to 18.5 mm and 4.9 mm, respectively. It is necessary to keep in mind that the competition between two neighboring roots for a given nutrient begins, when their depletion zones overlap. Even though, the question remains, which nutrient is dominant in this process, in turn impacting the whole plant metabolism? Solving this problem requires sets of data concerning root length density, which is variable both between crop species, stage of development and root's distribution in the soil profile. Therefore, K uptake characteristics of winter wheat and sugar beets, were compared at stages with maximum uptake rates, i.e., at heading and in the second half of July, respectively (Table 2). The calculated half distance between neighboring roots, as a result of root length density variability, increased with the soil profile depth. The degree of nitrate's utilization by wheat was the highest in the top-soil, several times exceeding its potential to deliver the required amount of nitrogen. The most spectacular is the fact that in the 7th-day period, the depletion of the nitrate-nitrogen zone extended down to 150 cm. At the same time, the K depletion zone occurred at the depth of 60 cm. In the case of sugar beet, the depletion zone for nitrates reached down to 90 cm, whereas for potassium only to 30 cm.

Depth	Winter wheat [1]				Sugar beet [2]			
	L_{vr}	r_1	Dd, %		L_{vr}	r_1	Dd, %	
	$cm \cdot cm^{-3}$	cm	NO_3^-	K^+	$cm \cdot cm^{-3}$	cm	NO_3^-	K^+
0-10					2.7	0.73	509	134
10-20	8.2	0.42	888	234	3.1	0.68	546	144
20-30					2.1	0.83	449	118
30-60	1.7	0.92	404	106	0.38	1.95	191	50
60-90	1.0	1.20	310	82	0.26	2.35	158	42
90-120	0.7	1.43	259	68	0.08	4.24	88	23
120-150	0.27	2.31	161	42	0.01	12.00	31	8
150-190	0.03	6.93	54	14	-	-	-	-

Adapted from [1][36] and [2][28]

Table 2. Effect of root length density distribution in the soil profile on the degree of potassium and nitrate-nitrogen depletion at the critical stage of potassium accumulation by two crops

These two sets of data outline some important information concerning the management of both nutrients. Firstly, at the critical period of each crop development, nitrogen should be considered as a nutrient significantly limiting plant growth. In the case of wheat as a crop accumulating a significant amount of nitrogen in grain, an external supply of this nutrient is required at heading to fulfill this goal. A quite different strategy should be recommended for sugar beet, since it reaches at the critical stage of K accumulation maximum rates of both dry matter and nitrogen accumulation [26, 37]. This crop in subsequent stages of growth relies on soil N resources, which uptake is governed by K supply. The second information refers to un-depleted resources of potassium, present in deeper soil layers. These reserves can be considered as the basic source of K supply during critical stages of beet growth and/or during any kind of growth disturbance due to stress. It is worth mentioning, that water shortages first limit nutrient flow in the topsoil, and then extend down the soil profile. Therefore, soil K reserves present in deeper soil layers are important for the exploitation of the plant yielding potential or to protect its growth under stress.

The effective transformation of solar energy into plant biomass depends on the supply of nitrogen, which is crucial for both carbon fixation and its subsequent partitioning among plant organs. Therefore, the rate of plant growth, taking into account its aerial part, is determined by nitrogen availability, especially nitrate ions. It is recognized, that higher soil moisture content usually means greater availability of nutrients to plants. Nitrogen supply to roots is *via* the transpiration stream of water (mass flow). Processes leading to the decrease of soil water content are the main reasons increasing the importance of diffusion as the core mechanism of nutrient transportation towards roots [30, 33, 38]. Nitrogen fertilizer use by a crop is related to the soil K fertility level. It has been documented, that insufficient supply of K results in lower, than expected, uptake rate of nitrate-nitrogen, which in turn decreases the rate of aerial biomass growth. This specific phenomenon is explained by the fact, that potassium accumulated at the root surface controls nitrogen inflow into the root. The rate of nitrate's transport through roots into the shoot depends on K concentration in the soil solution, governed also by K soil fertility level. At the same time, malate is produced in the shoot and part of the K-malate undergoes recycling through the root system [39]. Therefore, external and internal K sources are responsible for effective uptake of nitrogen from its soil pool. Insufficient supply of potassium from the soil solution significantly restricts the uptake of nitrates, reducing in turn their concentration in the root and consequently their transportation into leaves, where they undergo reduction. This also means, that the plant is not able to take up adequate amounts of N, when K is in limited supply. It can be concluded, that high-yielding crops require excessive supply of K in order to match their demand for N in critical stages of yield component's development.

3. Potassium as a water-stress ameliorative agent

3.1. Plant growth stages – Yield forming function of potassium

Yield can be defined as the end-product of yield component's expression of a particular crop during growth and development. According to the concept *multiple limitation hypothesis* [MLH, 17], water and nitrogen supply plays a decisive role in assimilates partitioning among main crop plant organs [40, 41, 42]. These two nutrients affect both the rate of dry matter accumulation and yield component's development. In order to understand their influence on the rate of dry matter accumulation of the yield, the whole life cycle of a crop can be divided into three main periods: i) yield foundation (YFP), ii) yield construction (YCP), iii) yield realization (YRP) [43]. The shortage of potassium can affect plant growth in each of the above indicated periods. The key problem remains, about how potassium improves yield forming effect of water and nitrogen?

Potassium concentration in plant biomass varies from 1 to 5% of dry matter weight. When soil potassium is deficient, plant growth is reduced, resulting in smaller, dull bluish-green and wavy leaves and thinner stems. Plants often tend to wilt. Visual symptoms of potassium deficiency are easily recognized, but only those under severe potassium shortage, (Photo 1). In early stages of growth and also under hidden K deficiency, its visible symptoms are eventually confused with nitrogen deficiency. The main reason is the slow rate of the aerial biomass growth. The shortage of potassium is the reason of basic physiological processes disturbance, which in turn negatively affect the development of yield components. Potassium deficient plants: a) develop a weak root system, b) are not efficient in nitrogen uptake, c) grow slowly, d) develop infirm stems and lodges frequently, e) use inefficiently water and nitrogen, f) show high susceptibility to diseases, g) yield miserably, h) produce lower quality yield [19, 20]. The recent study conducted in Canada showed that the shortage of potassium reduces grain of maize by 13% [44]. In the Central-Eastern European countries, during the last two decades, the supply of nitrogen has not been balanced with potassium and phosphorus, in turn seriously limiting harvested yields of cereals [45].

In the yield foundation period (YFP), K supply affects both the root system and aerial parts build up. In cereals, it stops at the end of tillering, and in all dicotyledonous crops, at the rosette stage. In general, at the beginning of the plant life cycle, the supply of water and nutrient is not considered as the factor limiting the root system extension. As a rule, all nutrients are uniformly distributed, and roots follow the genetically fixed patterns [30, 46]. Plants grown in soil fertilized with potassium, i.e., in the K fertile soil, show at early stages much higher rate of root system ingrowths. As a result, roots of plants well supplied with K are able to reach the deeper soil layers considerably earlier, than those poorly K-nourished. For example, the daily rate of extension of sugar beet roots, due to ample K supply can be accelerated by 50% as compared to plants grown in the K medium K level, irrespectively of the weather course. The same degree of maize response to high K availability has been documented. Cereals, for instance spring barley showed a much weaker response to the elevated K soil level [47]. The observed phenomenon supports the hypothesis, that K induces adaptation of crops to summer semi-drought, which emerges irregularly in temperate regions.

The key attribute of the yield construction period (YCP) is the linear rate of the dry matter increase. At the end of this period, crop plants reach the highest rate of growth. Therefore, K supply during this particular period is considered as the critical factor of yield performance. In cereals, it extends from the end of tillering up to the end of heading [48]. In other seed crops, the most sensitive phase to K shortage extends from the rosette up to the budding. Vegetable crops show sensitivity to K supply from the rosette up to technological maturity. For high-yielding crops, K supply is crucial for maximizing the dry matter accumulation and critical for yield component's development (Table 3). For example in maize, the shortage of potassium during anthesis negatively affects the number of kernels on the cob [49, 50]. As shown in Table 4, plants grown on light soil showed poor development of this yield component, mainly due to the extremely strong response to drought in 2006. Therefore, any shortage of K during the linear period of each plant growth is considered as critical for final yield development.

The yield realization period (YRP) of a particular crop extends from the beginning of anthesis up to final maturity. The shortage of K affects the most vegetable crops, including potatoes. Seed crops are also sensitive to K supply during ripening, especially in regions of year-to-year weather variability. For example, the content of potassium in the flag leaf of winter wheat at the stage of milk grain maturity can significantly affect the yield (Y), as presented by the equation [51]:

$$Y = 2.35K + 4.0; R^2 = 0.75 \; n = 9 \text{ and } P \leq 0.01 \tag{7}$$

Crop	Visual symptoms of deficiency	negatively responded component/parts
Cereals	dull green color; tip and marginal chlorosis on lower leaves	- number of ears per unit area, - grains per ear
Maize	curling of leaves by mid-morning, tip and marginal chlorosis on lower leaves	- number of kernel's rows, - number of kernels per row
Oil-seed rape	smaller rate of rosette growth, tan-coloring of lower leaves	- secondary branches, - capsules per branch
Sugar beet	early leaves wilting during heat of the mid-day, tip and marginal chlorosis on older leaves	- size of leaves, - weight of storage root, - content of sugar
Potato	wilting-like shape of canopy during heat of the mid-day, tip and marginal chlorosis on older leaves	- number of stems, - weight of tubers, - content of starch

Table 3. Effect of potassium deficiency during the linear period of crops growth on visual symptoms and yield components development

Experimental factor	Level of factor	Yield and elements of yield structure							
		Yield, t ha[-1]		NKR[3]		NKC		TGW, g	
		L[1]	M[2]	L	M	L	M	L	M
Fertilizing treatments	NP	6.49[a]	8.27[a]	24.9[a]	26.8[a]	357[a]	390[a]	239[a]	259[a]
	NPK	7.05[b]	9.61[b]	26.3[a]	29.2[b]	376[a]	421[b]	251[ab]	284[b]
	NPKMgS	7.46[b]	10.1[b]	26.3[a]	29.1[b]	370[ab]	411[ab]	261[b]	306[c]
	NPKMgSNa	7.35[b]	10.6[c]	24.6[a]	27.7[ab]	365[ab]	405[ab]	249[ab]	323[d]
Years	2005	8.85[b]	10.3[b]	31.7[b]	27.0[a]	473[b]	410[b]	275[c]	301[b]
	2006	2.73[a]	8.13[a]	14.3[a]	25.7[a]	159[a]	334[a]	217[a]	298[b]
	2007	9.69[c]	10.5[b]	30.6[b]	31.9[b]	470[b]	476[c]	259[b]	277[a]

Source [55]

[a]means with the same letter are not significantly different at α=0.05 (Tukey test);

[1]L: light soil - loamy sand, [2]M: medium soil - sandy loam; [2]NR - number of rows per cob,

NKR – number of kernels per row, NKC – number of kernels per cob, TGW – thousand grain weight.

Table 4. Statistical evaluation of main factors affecting yield and structural components of maize grain yield at the background of soils differing in texture

This finding corroborates the importance of the subsoil K reserves for efficient management of N, as the nutrient decisive for leaves activity during ripening of cereals. The grain weight increase in response to K supply is probably related to its effect on assimilates transportation in the phloem [20]. Thousand-grain weight (TGW), a structural parameter of a grain yield of seed crops, indirectly describes a plant nutritional status in this period. The final weight of kernels generally reflects the crop canopy capability both to produce and to supply carbohydrates to growing kernels [52]. As presented in Table 4, this yield component showed a significant response to all studied factors, but the soil complex was the most important. Plants grown on soil, naturally reach in potassium, achieved TGW by 17% higher as compared to those grown on light soil, in spite of the same content of K initially available.

Water requirements of plant crops are variable accordingly to the stage of their growth. The most sensitive stages cover the linear phase of biomass accumulation [see equation No. 7]. Sugar beet and potato plants are responsive to water supply during the most of the season, but especially during the highest rate of the dry matter increase, i.e., in the mid-season (July and August in the temperate regions of the world). Soil water capacity is a function of its textural class and precipitation over the whole season. It has been documented, that long-term fertilization with potassium results in increasing content of plant available water [47]. This phenomenon is probably explained by specific, glue-like action of potassium ions to individual soil grains [53]. The spatial pattern of water uptake from various regions of the soil profile depends on both soil moisture and roots distribution [46]. Water uptake and extraction patterns are related to rooting density. For example, a high-yielding winter wheat extracts 50 to 60% of total water from the first 0.3 m; 20 to 25% from the second 0.3 m; 10 to 15% from the third 0.3 m and less than 10% from the fourth 0.3 m soil depth. The usability of water by plant root from deeper layers depends on its penetration ability [54]. However, the deepest parts of the soil profile are responsible for water and nitrogen supply during stages of maximum dry matter accumulation.

The maximum rate of water use by crop occurs at field capacity, i.e. at maximum soil available water content. As the soil dries, the attainable soil water content decreases, leading to a significant drop in plant water potential, which also depends on plant structure and transpiration rate. At the onset and during sustained periods of drought, highly synchronized responses occur between root and shoot tissues. Signals from the roots have almost immediate effects upon shoot growth and its physiological functions, which modify the crop plant response to drought, in turn its productivity. The prolonged drought disturbs the diurnal rhythm of stomata, which are not able to control water loss from the leaves, further increasing the stress. Next, photosynthesis rate declines and respiration tend to increase, reducing consequently, dry matter accumulation. Shortage of assimilates transport to roots decreases the rate of their growth and as a consequence root system may be less able to utilize reserves of water stored in deeper soil layers [38, 46, 56].

The main agronomic problem, but not only, is the question how the water deficit may be ameliorated? In agriculture practice irrigation and breeding used to be treated as the main ways for overcoming water shortages. The simplest solution is to supply more water, i.e., to irrigate. However, not all farmers can invest in irrigation equipments. The second solution is to find out varieties, well adapted to water shortage. So far, in spite of huge investigation, breeding for drought resistance, remains still the open-box [56, 57]. It is well known, that root morphology is guided genetically, but the ultimate shape of the root system largely depends on the effects of environmental factors. The depth of the soil reservoir that holds water available to a plant is, in fact, determined by plant's rooting, in turn depending on soil characteristics, including compacted layers and water storage. Hence, the extension of roots into deep soil layers is crucial for crop performance under limited water supply. Drought adapted plants are characterized by great and vigorous root systems [58]. Experimental studies conducted in England showed that winter wheat roots below 1 m contribute only to 3% of total root system weight, but at the same time it delivered 20% of the transpired water during dry periods [54].

Under field conditions, water availability and its supply to currently growing crops is year-to-year variable, in turn affecting seasonal yields variability. Therefore, yields harvested by farmers in *good years*, i.e., under relatively ample supply of water, are usually higher, expressing higher unit productivity of the applied nitrogen and *vice versa*. It is recognized, that plant growth is better maintained under stress if adequate amounts of nutrients are available throughout the growing season. The deficit of nutrients reduces the rate of metabolic processes in the plant, making energy transfer and other growth processes less efficient. The adequate, balanced supply of N, P, and K should meet crop requirements, keeping its healthy and vigorously throughout the growth season [19, 30, 33]. This conclusion is corroborated by data presented in Table 4. In 2005 and 2007, favorable for maize growth, yields of plants fertilized with NPK were significantly higher compared with those fertilized only with NP, irrespectively on soil texture. In the extremely dry 2006 year, in spite of the same input of fertilizers and soil K fertility level, harvested yields were by $2/3$ and $1/4$ lower as compared to good years, respectively for loamy sand and sandy loam. Grain yield responded to applied nutrients, but it was non-significant on the light soil (Fig. 5). This example implicitly indicates on the importance of inherent soil K fertility in ameliorating water shortage, significantly affecting crop growth during the Yield Foundation Period (Table 4). It can be therefore concluded that on light soils (L), K application ameliorates mild but not severe stresses. Soil originated from loams (medium soils, M) are much more resistant to drought, allowing to take under control water stress, provided a well potassium management.

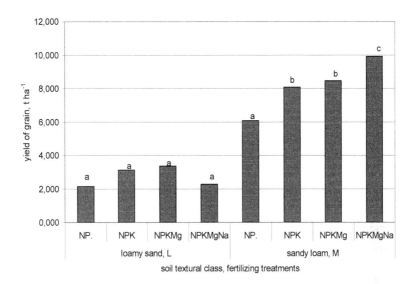

Figure 5. Response of maize grown on soil differing in texture to increased fertilization level in 2006 (dry year); Source [55]

3.2. Impact of potassium on WUE – Maize as a case study

The water-management index describing water-use efficiency (WUE) refers to the quantity of biomass produced by a crop plant per volume of water transpired and evaporated during its life cycle. In agronomy, the WUE index termed as the crop water productivity (CWP) relates the quantity of actually harvestable or marketable crop plant part (seeds, grain, roots, tubers, etc.) produced on a given area in a fixed period of time (yield, Y) per unit of transpired water [59]:

$$CWP = Y_a / ET_a \tag{8}$$

The actual evapotranspiration (ET_a) defines the amount of water use (transpired and evaporated water) (mm, m^3) by the cultivated crop during its growth period. Its value is constant for a particular crop within a given geographical region, in spite of a slight year-to-year variability. For example, indices of ET_a calculated for maize in two contrastive growth seasons

2006 and 2007, with respect to the weather course, amounted to 622 and 572 m^3, respectively (see Table 4). However, in 2006, harvested yields on light soil were much lower than on the medium one (Fig. 5). In contrast in 2007 yields were both considerably higher and did not show dependence on soil texture. Therefore, the applicability of the CWP index for evaluating yielding effects of agronomic factors is limited.

The French and Schulz approach, expressed as the water limited yield concept (WLYC), [60] is proposed [60] for the description of the impact of K on water management. The algorithm for the water limited yield (WLY) calculation is as follows:

$$WLY = TE\left(R + WR - \Sigma E_s\right) \tag{9}$$

where: TE, maximum unit water productivity, fixed at the level of 20 kg ha^{-1} mm^{-1}, R refers to the sum of rainfall during the growth period, WR expresses water reserves in the soil profile down to 1 m, and ΣE_s, represents the seasonal soil evaporation, equals to 110 mm.

The proposed procedure takes into account two variables affecting WUE, resulting in yield fractionation. The first yield fraction (WLY), reflects a maximum yield at a given amount of attainable water to a crop during its growth. Controversies about the applicability of the Eq. No 9 refer mostly to the threshold value of the TE, which was originally set up for wheat at the level of 20 kg of grain per 1 mm of water [61]. In maize, taking into account its higher water-use efficiency, this threshold value is questionable and should be fixed at a slightly higher level. The another controversy refers to the importance of water reserves, WR, present in the soil profile. This water reservoir is responsible for both water, and nutrients supply at early stages of a plant growth. Therefore, this component of soil water characteristic has been introduced by Authors (the current chapter) into the original French and Schulz equation. The second yield fraction quantifies the net effect of the applied agronomic measure on WUE, resulting in yield gain or loss.

The graphical interpretation of the WLY concept, as proposed by Authors, allows to discriminate the effects resulting from the action of transpired water and that of the tested factor. As shown in Fig. 6, the maximum yield of maize was higher in the favorable year 2001 as compared to the dry one, i.e., 2003. The effect of increasing nitrogen rates was dependent on potassium management. In the treatment without K application, the highest yield increase due to N was documented for its rate of 100 kg ha^{-1}. In contrast, on plots with current K application, the highest yields were harvested in the treatment with 140 kg N ha^{-1}, irrespectively of the season. The relative contribution of K application in the final yield, measured for this particular treatment, was 40% and 6% in 2001 and 2003, respectively. It can be therefore concluded, that the exploitation of maize potential significantly depends on the nitrogen rate, but adjusted for the K fertility level. Therefore, any inadequately recommended N rate can result in yield decrease in good years or even its depression in years with drought, as occurred in 2003 (Fig. 6) and in 2006 on the light soil (Fig. 5).

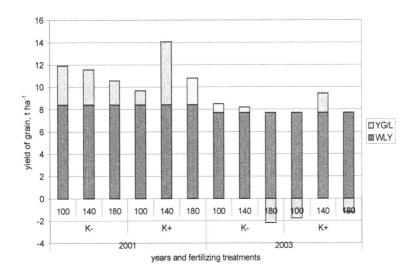

Legend: WLY – water limited yield; YG/YL – yield loss/gain, K-, K+ - K fertilizing treatments; 100, 140, 180 kg N ha⁻¹.

Figure 6. Effect of soil K fertility level on maize yield in two weather contrastive years; Adapted from [62].

4. Soil potassium resources – K availability to crop plants

4.1. Soil K mining

As described in the first part, the consumption of K fertilizers in many parts of the world has significantly decreased. The annual shortage of K in the global scale is calculated at the level of 20 kg ha⁻¹ [63]. Decades of cropping without sufficient replacement of K removed by harvested plant portions depleted soil K resources to the yield-limiting level. The long-lasting negative K balance is nowadays considered as the second factor of agricultural soil productivity degradation, following soil acidity. On the average, 18.6% world soils is extremely poor in potassium. The worst situation occurs in South-East Asia (43.5%), followed by Latin America (39.3), Sub-Saharan Africa (29.7%), East Asia (19.8%) [64]. Central Europe and countries originated from Former Soviet Union are also threatened by soil mining, because 25% of arable soils present low content of potassium [65].

A minimum of 300 kg ha⁻¹ of available potassium is required for a good growth of high-yielding crops, assuming 33% of its utilization by crop [66]. In low-input systems, crop production mostly relies on soil resources and alternative sources of nutrients, including K mineral fertilizers. The high-input systems, which do not cover, or even replenish plant K needs at critical stages of yield development, result in soil K mining. High year-to-year variability and/or yields stagnation is not always recognized as the attribute of the inappropriate K management. Therefore, all potassium mined soils as well as light textured and also organic soils should be considered as risky for crop production. For all these groups, recommended rates of applied K should be greater than its removal.

4.2. Soil K pools

The total content of soil potassium in the top-soil (layer 0-0.2 m) ranges, depending on soil texture from ca 1 000 to 50 000 kg K ha^{-1} [67]. Therefore, it can be concluded, that whole reserves of K in the rooted soil profile (down to 1.0 m) are several times larger. However, most of the soil potassium is not directly attainable for currently growing crop. Soil K resources are distributed in pools, which release K$^+$ ions with different rates, depending on geochemical characteristics of a particular pool. Based on chemical extraction procedures and probability of K uptake by a meanwhile grown crop, four operational K pools/forms have been defined: i) water-soluble (WSK) ii) exchangeable (EXK), iii) non-exchangeable (NEXK, iv) structural/mineral (MIK). The first one, containing K$^+$ ions present in the soil solution, is directly available to the plant. In Polish soils, it content ranges from about 60 to 90 kg K ha^{-1} for the light and heavy soil, respectively (Fig. 7). This form of potassium is at its highest level in spring and decreases throughout the growth season as plant takes it up. It covers plant needs at early stages of growth, but not in the high-season. This K pool is also sensitive to leaching, which in temperate regions of the world takes place in autumn and winter, provided water saturation of the whole soil profile. The amount of leached K is inversely related to soil texture, ranging from 1 to 8 kg K ha^{-1} for soil originated from loams and sands, respectively [66].

The second K pool (EXK) contains K$^+$ ions held by negatively charged clay and humus particles. In Polish soils, the amount of the EXK ranges from about 200 to 650 kg K ha^{-1}, for very light and heavy soils, respectively. For this K form a threshold content is fixed at the level of 100 mg kg^{-1} [67], i.e., 360 kg K$_2$O ha^{-1}. The first two K pools are in a dynamic equilibrium, enforced by the presence of the plant root. According to the Le Chatelier-Braun principle of *contrariness*, any changes in K$^+$ ions concentration in the soil solution results in their movement from the exchangeable to the soil solution pool. The reverse process occurs in response to K fertilizer's application. Both pools, when not replenished with K in fertilizers or manures, undergo depletion, decreasing the capacity to match plant demand in time and space [68, 69]. Under lack and/or insufficient K delivery from external sources to currently grown crop, which even in the high cropping systems is not exception, but a rule, its growth and productivity depends on the non-exchangeable soil resources (NEXK). This pool is several times larger than the EXK one, as shown in Fig. 7. For this K form, the threshold level Is fixed at 400 mg kg^{-1} [70], i.e., 1440 kg K$_2$O ha^{-1}. The fourth pool (MIK) represents K in soil rocks and minerals. This pool is considered as long-term K reservoir, highly dependent on the type and the weathering rate of K bearing minerals [Table 5].

Minerals [1]	Formula	g • kg[-1]	Rocks [2]	g • kg[-1]
K feldspar	$KAlSi_3O_8$	140.3	Sanstone	12.3
Leucite	$KAlSi_2O_6$	178.9	Clays	23.3
Nepheline	$(Na,K)AlSiO_4$	130.0	Shales	20.4
Kalsilite	$KAlSiO_4$	246.8	Limestons	2.6
Muscovite	$KAl_3Si_3O_{10}(OH)_2$	90.3	Chernozem[3]	8.4-22.0
Biotite	$K_2Fe_6Si_6Al_2O_{20}(OH)_4$	76.2	Cambisols[3]	11.4-20.9
Phlogopite	$K_2Mg_6Si_6Al_2O_{20}(OH)_4$	93.8	Vertisols[3]	16-28.5

Source [1][72], [2][73], [3][74]

Table 5. Potassium content in K-bearing minerals[1], rocks[2] and soil[3]

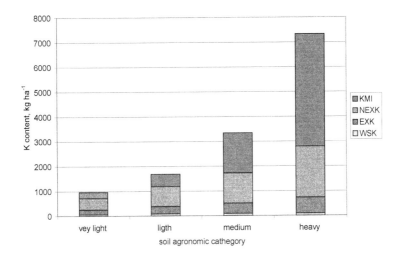

Figure 7. Distribution of potassium among pools in Polish soils at the background of soil texture; Adapted from [71]

4.3. Factors affecting potential availability of the NEX-K to crops

In the majority of cropping systems, harvested yields depend on the non-exchangeable K pool [68, 69]. The yielding impact of this K form increases in most systems, where K removal by crop is not fully replenished. In order to elucidate the importance of this K pool for plant production, an example of four different fertilizing systems on spring barley yields is short-ly described. The status of K forms in black earth after twelve cycles of three-course rotation, as presented in Table 6, showed a significant decrease in the content of both available and slow-released K forms. The EXK pool was much below the standards (100 mg K kg soils[-1]), irrespective of the fertilizing system. The NEXK was several times larger, exceeding the threshold level in three of four treatments, i.e., in K fertilized ones. The described study implicitly revealed the significant effect of the NEXK pool on yield of spring barley (Y):

$$Y = -0.456 + 0.008NEXK; R^2 = 0.69; n = 16. \qquad (10)$$

In Poland, the official recommendation for K is based on the Egner-Riehm extraction procedure (Doppel-Lactat, pH 3.55). The harvested yield of barley grain also showed a significant dependence on the content of available K (ERK), following the quadratic model:

$$Y = -0.84 + 0.107ERK - 0.00048 ERK^2; R^2 = 0.55; n = 16 \qquad (11)$$

This type of relationship between yield and available K means, that potassium supply limited the yield of grain to a certain value, which in this particular case was fixed at 111.5 g K kg[-1] soil. This value implicitly indicates the FYM treatment as optimal for the maximum yield of barley. In the third step of evaluating the yield forming effect of K, the ERK was regressed against K content in other K pools. The applied stepwise regression implicitly revealed its significant dependence on the NEXK, The reliability of the ERK pool prediction was improved by implementing the EXK into the model:

$$ERK = -5.7 + 0.157NEXK; R^2 = 0.51 \text{ and } P \leq 0.001 \qquad (12)$$

$$ERK = -17.9 + 0.12EXK + 0.12NEXK; R^2 = 0.71 \text{ and } P \leq 0.001 \qquad (13)$$

Potassium treatments	K pools				Egner-Riehm K ERK
	WSK	EXK	NEXK	MIK	
	mg kg[-1]				
Control	7	17	378	1045	50
NPK	16	33	648	949	84
FYM	19	28	695	1140	114
1/2NPK + ½FYM	10	33	565	855	61
Source [68]					

Table 6. Effect of 36 years of continuous fertilizing systems on the distribution of potassium forms

The efficient use, i.e., exploitation of the non-exchangeable K pool in crop production requires to use specific agronomic methods. The most farming efforts are focused on increasing both amounts of plant available potassium and crop accessibility to this particular soil pool. There are numerous processes involved in the equilibrium between exchangeable potassium (EXK) and non-exchangeable potassium (NEXK) pools. The basic way in reaching both goals simultaneously is to fix the soil pH at a level adequate for the most sensitive crop in the crop rotation system. The application of lime induces a series of interrelated processes, resulting in the improvement of fundamental growth conditions for crop plants. Therefore, their demand for nutrients, including potassium, increases proportionately. Aluminum (Al^{3+}) neutralization is the primary effect of lime application, which in turn creates a chemical and physical milieu for better growth of roots. This action is the key agronomic practice responsible for increasing the accessibility of a given crop to potassium resources in the soil profile. Other processes induced by lime results in increasing amounts of available potassium in the soil solution. The key one is directly related to the disturbance of the K^+/Ca^{2+} equilibrium at the interface soil solution/EXK pool. The sudden increase of Ca^{2+} ions concentration in the ambient soil solution is attributed to the accelerated rate of K^+ displacement from the cation exchange capacity (CEC). The another consequence of liming is the proliferation of soil fauna, which increases the rate of organic matter decomposition. The induction of microorganisms activity results in series of secondary processes affecting:

a. the release of K from organic matter,

b. the displacement of K^+ ions from both the EXK and NEXK pools,

c. the build-up of soil CEC,

d. the dissolution of non-exchangeable K from clay particles.

The processes reported in positions *a* and *d* are of a great importance for the current and long-term soil K economy, respectively. However, both are efficient the most under conditions of a slight acid pH. At neutral pH, the elevated concentration of Ca^{2+} slows down the effect of H^+ on K^+ displacement. With respect to the third process (position *c*), the build-up of soil organic matter content, oriented on increasing soil CEC is the long-term strategy of K management. The increased size of CEC should be considered as the extended reservoir for cations, both potentially threatened by leaching from the soil solution and/or dissolved from the non-exchangeable K pool.

In the last decade, a lot of scientific projects dealt with microorganisms, considered as a tool for increasing the availability of K from non-exchangeable potassium (NEXK) and that occluded in rocks and minerals (MIK) pools. The study carried out with plant growth-promoting Rhizobacteriaceae (PGPR) showed, that some bacterial strains such as *Bacillus edaphicus*, *Bacillus mucilaginosus* are capable to release potassium from silicate minerals. Their action in K-bearing minerals is *via* H^+ ions, and/or by organic acids (citric, tartaric, oxalic), active in divalent cations complexion [72, 75, 76, 77]. A similar effect is expected when plants such as cotton, grasses, legumes, crucifers were used.

The importance of externally incorporated microorganisms to arable soil in raising up soil fertility as described above was mostly limited to laboratory experiments. The study conducted with the application of bio-fertilizers in Poland showed, in general, a significant increase of mineral nitrogen content, indirectly stressing on the accelerated rate of organic matter decomposition. Much larger amounts of released nitrates in response to increasing fertilizers application and bio-fertilizer indicate an efficient rate of ammonia nitrification, which in turn generates H^+. The formulated hypothesis assumes a local soil acidification, which results in a significant increase of cations and phosphorus contents. The highest increase of the latter ones suggests a multifunctional action of soil applied microorganisms (Table 7).

Treatments	$N-NH_4^+$	$N-NO_3^-$	P_2O_5	K_2O	Mg
	kg ha^{-1}		mg kg soil^{-1}		
N	24.3 [1]	55.1 [1]	36.2	91.6	76.6
N + biofertilizer	26.6	72.1	59.2	111.7	97.0
NPK + biofertilizer	28.4	107.4	91.5	140.4	88.9

Source [78], [1]extracted in 0.01 M CaCl$_2$,

Table 7. Effect of a bio-fertilizer on the post-harvest content of available nutrients in soil cropped with potato[1]

5. Crop rotation – The background of soil fertility management

Crop rotation describes a sequence of crop plant species cultivated on the same field within a fixed time. Three classical principles of crop rotation include: i) an appropriate choice of cultivated species, ii) crop frequency, taking into account some biological limitation, iii) fixed crop sequence. Crop rotation, in fact, under a particular climate and soil agronomic properties of a field, defines the structure and management of applied inputs. The main goals of crop rotation are:

a.   yield stability, as a basis of a long-term stabilization of farm economy,

b.   amelioration of the resistance of growing plants to stress, mostly of biological origin,

c.   optimization of the use of soil resources, with respect to water and nutrients.

All these goals were rigorously guarded by farmers up to the end of the first half of the XX century. The technical progress, which started at the beginning, but accelerated in the second half of the XX century, resulted in a great increase of agriculture means of production, including fertilizers and pesticides. In addition, for the period extending from 1950 to 1970, intensive breeding programs within the "Green Revolution" succeeded first in new wheat and rice varieties, and next other cereals and maize. The main attribute of high-yielding varieties at that time was the extended capacity to accumulate nitrogen. Soon, the rigid crop sequence rules, based on legumes as a source of nitrogen and assuring its biological protection became limiting factors in a sharp yield increase. Consequently, changes in sequences of cultivated crops practiced by farmers were more and more oriented on net income, neglecting at the same time the efficiency of applied nitrogenous fertilizers, pesticides. This in turn increased the pressure of agriculture on environment [2, 80, 81].

New paradigms of agriculture development, oriented on sustainable use of resources, can be achieved, provided crop rotation rules are introduced. The modern view on principles of crop rotation arrangement takes into account its flexibility in the selection of crop species, depending on market needs. Nowadays, the objectives of the rational crop sequence should strictly consider i) the farm economic profitability – its adaptability to market oriented changes, ii) the optimization of resource use, both internal (soil) and external (fertilizers, pesticides), iii) the minimization of the impact of agriculture on local and global environment (nitrogen, phosphorus) [82, 83]. Therefore, a profitable crop production requires the development of alternative strategies, oriented on a well-thought-out management of water and nutrient resources in a particular crop rotation.

The reported expectations and assumptions regarding crop rotation refer also to potassium management. There are some experimental data supporting the concept of sustainable use of soil potassium, based on crop rotation principles:

a. soil, taking into account the whole profile, must be sufficiently reach in K to supply sufficient amount of potassium to a high-yielding crop within an extremely short period of growth – the critical period of yield components formation, to assure maximization of its yielding potential exploration [21, 24],

b. leafy crops, for example sugar beet, oilseed rape, to cover K requirements during crucial stages of growth, need to explore a thick layer of the soil profile [28, 84],

c. root system of leafy crops is much weaker in comparison to cereals (Table 8), being a prerequisite of higher level of available potassium,

d. demand of cereals for potassium is much lower than leafy crops; root density is at the same time much higher, hence a higher efficiency in K uptake (Fig. 8; Table 8),

e. both leafy crops and cereals respond more to soil fertility K level than to freshly applied fertilizer K [27, 66, 85, 86],

f. all crops respond to current potassium fertilization in years with stress, mostly related to water shortage and site specific diversification of K management [47, 55, 86, 87].

Crops	RL_V, cm cm^{-3}	R_d, cm
Been	0.5-2.0	0.50
Oil-seed rape	-	1.00
Potato	1-2	0.50
Sugar beet	1-2	1.00
Barley	3-4	0.50
Wheat	3-8	0.80
Rye	4-8	1.00
Source [88], [89]		

Table 8. Root length density (RL_V)[1] in the top-soil layer and mean rooting depth R_d [2]

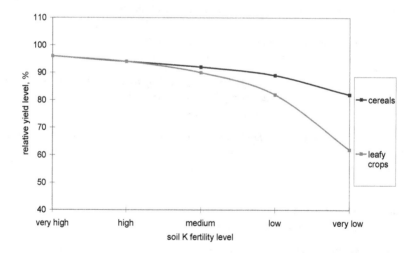

Figure 8. Effect of soil potassium fertility level on yield level of two groups of crop plants; Adapted from [90]

6. Crop rotation potassium balance – A strategic scale of K management

6.1. System of potassium fertilization – Key assumptions

The efficient system of potassium management should focus on requirements of the most sensitive crop in a particular crop rotation. The mandatory objective of effective strategy of any crops fertilization with potassium is to cover K demands of the currently growing crop during the period of their highest growth rate. However, crop plants grown in a fixed crop rotation present different sensitivity to current level of attainable soil K. Thereby, the primary objective of rational K management should focus on fulfilling the requirements of the most sensitive crop within the given crop sequence. It demands for K determines the top level of the critical range of soil available K for the whole sequence of growing crops. The basic target is the non-limited supply of soil potassium to plants during stages of the highest biomass increase, which coincides with the end of the linear period of dry matter accumulation. The degree of the K requirement covering by the most sensitive crop is decisive for both: i) water-use efficiency, ii) nitrogen utilization use efficiency. Any increase of these two indices results in the degree of yield component's development, considered as a crucial for a yield increase. The secondary objective is to select production measures essential for reaching the required level of available K [66].

In general, the crop-oriented K fertilizing strategy relies on the view that crops such as sugar beet, potato, oil-seed rape, (grain) legumes are significantly more sensitive to K supply than cereals, when grown subsequently in a fixed crop rotation. Therefore, the economically successful and environmentally sound K fertilization system should based on five pillars, assuming that leafy crops:

a.   are grown in rotation with other crops, mostly with cereals,

b.   have a substantially weaker root system as compared to cereals,

c.   express considerably higher quantitative requirements for K at critical stages,

d.   can explore a thick volume of the soil profile,

e.   are, in consequence, much more than cereals sensitive to the level of available soil K.

The crop rotation-oriented strategy of K management also assumes a maximized recycling of internal, i.e., soil available K sources and field resources (plant residues). Therefore, the amounts of fertilizer K needed to cover its losses due to exports from the field in harvested products or/and leaching processes, depend also on the management of crop by-product (residues). The mentioned concept is in accordance with the Ideal Soil fertility (ISF) approach [91]. Potassium timing seems to be of secondary importance taking into account the strategic goals of K management. Natural growth conditions, mostly related to stressing factors, can only modify potassium fertilizer timing.

6.2. Potassium balance sheet in crop rotation – An operational procedure

Following the theoretical assumptions it appears that the main problem of adequate fertilization of the leafy crop with potassium is to develop an appropriate system of managing the potassium rotation-oriented system. The principal farmer's question is, how to achieve the target K availability range? The efficient K system development may consist of three basic steps:

a.   preparation of the K balance sheet for all crops in the fixed cropping sequence,

b. determination of the current level of available K,

c. correction of the current K level.

In practice, the balance sheet operates on the equation, which in a simple way quantifies K processes occurring at the field level:

$$K_{up} + K_{plr} + K_{nl} = K_{ni} + K_{fym} + K_f \quad \left[kg\ K_2O \bullet ha^{-1} \right] \tag{14}$$

Rearrangement of the equation No. 14 allows to calculate the potassium application rate:

$$K_f = \left[P_{up} \pm P_{plr} + K_{nl} \right] - \left[K_{ni} + K_{fym} \right] \pm dK_{av} \quad \left[kg\ K_2O \bullet ha^{-1} \right] \tag{15}$$

where:

K_{up} - K uptake by the main yield, kg • ha^{-1}

K_{plr} - K accumulation in plant residues, kg • ha^{-1}

K_{nl} - natural K losses (erosion, leaching), kg • ha^{-1}

K_{ni} - natural K input (dry and wet deposition), kg • ha^{-1}

K_{fym} - K supply in organic manure, kg • ha^{-1};

K_f - fertilizer K, kg • ha^{-1}

δK_{av} - intended/required change of soil available K, mg kg •soil^{-1}

In agronomic practice, some components of the K balance sheet such as natural input or losses may be omitted due to their minor importance as a source of K. The minimal set of data required when constructing the balance sheet for a particular crop sequence is as follows (Table 9):

a. unit K uptake (specific K uptake) by each crop cultivated in a given rotation, i.e., K accumulation in the main crop product unit and its respective amounts in the by-products, for example, in straw (expressed in kg K_2O • t^{-1} of the main product,

b. crop sequence in the fixed crop rotation,

c. methods of a specific management of by-products at the farm,

d. type of farm (crop, dairy, mixed) as related to manure production.

The first parameter shows a certain level of variability, according to soil, crop and production technology. The critical issue of the proposed concept relates to the management of the K_{plr} component of the Eq. No 15. All vegetative plant organs, such as straw or sugar beet tops, are very important sources of potassium. The environmentally and economically sound solution is to incorporate all harvested by-products into soil.

Components of balance sheet		Yield	Management of plant residues			
		t • ha^{-1}	exported from the field		left at the field	
			losses	input[4]	losses	input
Crop	products					
Sugar beet	storage roots[1]	60	60	-	60	-
	tops + root residues[2]	10 + 1	250 + 15[3]	13.5	250 + 15	238.5
Spring	grain	6.0	36	-	36	-
barley	straw + root residues	6.0 + 1.8	72 + 27	24.0	72 + 27	89.1
Oilseed	seeds	4.0	40	-	40	-
rape	straw+ root residues	10 + 2.8	200 + 42	37.8	200 + 42	217.8
Winter	grain	8	40	-	40	-
wheat	straw + root residues	9 + 2.55	126 + 38.3	34.4	126 + 38.3	147.9
Total		-	946.3	109.7	946.3	693.3
K net balance I		-	-836.6		-253	
Manure		34.0[5]	+214.4		0.0	
K net balance II		-	-622.4		-253	
K fertilizer needs, kg K$_2$O • ha^{-1}			622.2		253	
K fertilizer needs, kg K$_2$O • ha^{-1} • year^{-1}			155.6		63.25	

[1]main product, fresh weight;

[2]root residues + stubble - sugar beets ≈ 5%, cereals ≈ 15%, oil-seed rape ≈ 20% DW of aboveground biomass;

[3]content of K in plant residues;

[4]K recovery from plant residues/manure in the four-course rotation ≈ 90%; [5]fresh weight, K$_2$O content = 7 kg t^{-1}.

Table 9. An example of potassium balance, the 4-course rotation, kg K$_2$O • ha^{-1}

6.3. Determination of optimum soil K level

The efficient management of potassium in a given field depends on cultivated crop species and their cropping sequence. The main operational objective is to assess the degree of each crop sensitivity in the fixed rotation to the amount of soil available + fertilizer K. The graphical procedure of the optimum K range determination assumes, that the target crop shows the response to the applied potassium fertilizer, when soil K supply is too low to harvest 95% of the maximum yield. Based on data obtained from on-farm experiments and farmers experiences, it is possible to determine the perfect range of available K. As presented in Fig. 9, the applied statistical procedure, specifically the linear-plateau and quadratic regression models, allowed to fix the critical K point (limit), amounting to 170 mg K$_2$O kg^{-1}. However, as resulted from the analysis of the quadratic model, yield of the tested crop increased further up to 250 mg K$_2$O kg^{-1} (Fig. 9). In the case of sugar beets, the ideal level of soil K has been fixed at high level (clearly defined range), irrespectively of the site (soil) and year. All other leafy crops also require a fixed, in general, high level of soil available K during critical stages of yield formation. This level of soil attainable K content is the basis of the needed rate of K supply to a crop during important stages of yield formation. For cereals, the required K level is much lower. In general, 100 mg KEX kg soil^{-1} can be considered as the upper range of this plant response to soil K.

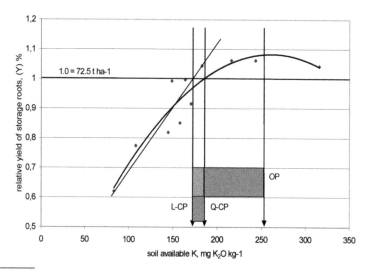

Legend: L-CP, Q-CP – critical point (quantity of available K) as determined by the linear (L) and quadrate (Q) regression models, respectively; OP- optimum content of available K.

Figure 9. A graphical method of assessing the critical available soil K for reaching maximum yield of storage roots by sugar beet; Adapted from [92]

The annual loss of potassium from the cropped soil due to intended export or leaching ranges from about 20 to 33%, depending on the crop [66]. Only in the case of leafy vegetable and fodder crops, its loss is substantially greater. Therefore, the required amount of potassium to be applied in rotation with leafy crops may be calculated using the equation No. 16. However, taking into account plant residues and their contribution to the required K amount, the needed quantity of purchased K fertilizer can be substantially, even by 3-times lower [Eq. No 17, Table 9]. Data concerning fertilizers value of crop residues can be obtained directly or calculated using constant relationships between K content in the main product and its concomitant by-product [66].

$$K_f = (K_{ar} - K_{ca}) \times 3 / 0.9 \ [K_2O \ ha^{-1}] \tag{16}$$

$$dK_g = K_f = [(K_{ar} - K_{ca}) \times 3 - K_{rec} / 0.9 \ [K_2O \ ha^{-1}] \tag{17}$$

where:

K_f – potassium fertilizer rate, kg K_2O ha^{-1}

K_{ar} - soil available K content required by the most sensitive crops in a fixed crop rotation, mg K_2O kg^{-1} soil, the critical range

K_{ca} - current level of soil available K, mg K_2O kg^{-1} soil

K_{rec} - K recycled in plant residues and/manure, kg K_2O ha^{-1}

3 - coefficient for converting soil K into K rates.

It is possible, based on specific K requirements to assign all cultivated crops into a particular soil available K classes. This has been reported in Table 10 for key crops cultivated in Poland. By using this operational scheme, the farmer can define a right place for crops grown in a particular crop sequence with respect to the required level of soil available K. This table can be considered as the first step in the development of the K fertilizing plan, oriented on K requirements of the most sensitive crop in the given crop sequence.

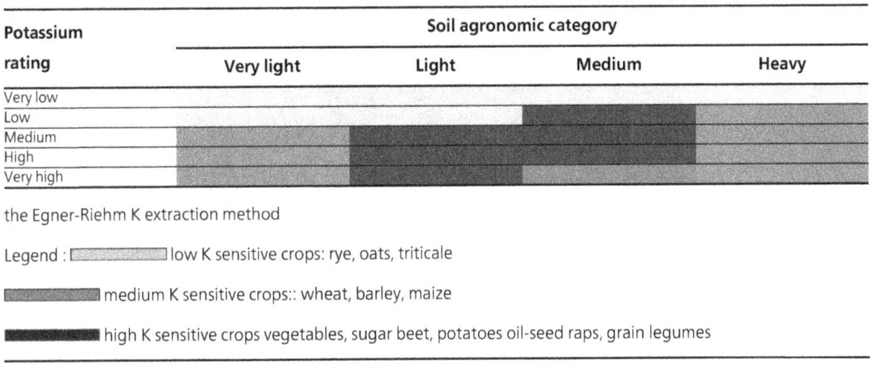

Potassium rating	Soil agronomic category			
	Very light	Light	Medium	Heavy
Very low				
Low				
Medium				
High				
Very high				

the Egner-Riehm K extraction method

Legend : ▭ low K sensitive crops: rye, oats, triticale

▬ medium K sensitive crops:: wheat, barley, maize

■ high K sensitive crops vegetables, sugar beet, potatoes oil-seed raps, grain legumes

Figure 10 Critical ranges of available potassium[1] for key crops in Poland

7. Conclusion

The modern nitrogen-oriented agriculture relies on mining of soil nutrient resources, mainly potassium. Therefore, its key attribute is low water- and nitrogen-use efficiency and high year-to-year variability of yields. Crop growth on soils poor in available potassium limits, consequently, food production in many regions of the world. Hence, the principal objectives of sustainable management of potassium are to: i) reduce year-to-year variability of harvested yields, ii) increase water- and N- use efficiency, iii) decreasing the pressure of agriculture on local and global environment. All applied agronomic measures should take into account K requirements of the most sensitive crop in a fixed crop rotation. Next, the required range of available K for the most sensitive crop, in turn should ensure effective N use during key stages of yield performance by all growing crops. The main way of fulfilling these goals is to gradually build-up or to re-build the attainable soil K pool. There are three key methods of economically profitable and environmentally sound K management. All of them rely on consecutive exploitation and use of: i) natural soil K reserves, ii) recycled organic K - plant residues, and in the last step, iii) K mineral fertilizers.

Photo 1. Potassium deficiency symptoms on maize; Author: Witold Grzebisz

Author details

Witold Grzebisz, Witold Szczepaniak, Jarosław Potarzycki and Remigiusz Łukowiak

Department of Agricultural Chemistry and Environmental Biogeochemistry, Poznan University of Life Sciences, Poland

References

[1] FAOSTAT. Available online at: http://faostat.fao.org/default.aspx. (accessed 2012-08-07).

[2] Spiertz JHJ. Nitrogen, sustainable agriculture and food security: a review. In: Licht-fouse E, Navarrete M, Debaeke Ph, Véronique S, Alberola C. (eds.) Sustainable agriculture. Dordrecht: Springer Science+Business Media BV; 2009. p635-651.

[3] Eickhout B, Bouwman AF, van Zeijts H. The role of nitrogen in world food production and environmental sustainability. Agriculture, Ecosystems and Environment 2006; 116 4-14.

[4] Galloway J, Cowling E. Reactive nitrogen and the World: 200 years of change. Ambio 2002; 30 (2) 64-71.

[5] Kirchmann H, Thorvaldson G. Challenging targets for future agriculture. European Journal of Agronomy 2000; 12 146-161.

[6] Tait J, Morris D. Sustainable development of agricultural systems: competing objectives and critical limits. Futures 2000; 32 247-260.

[7] Groot JJR, Penning de Vries FWT, Uithol PWJ. Food supply capacity study at global scale. Nutrient Cycling in Agroecosystems 1998; 50 181-189.

[8] Pretty JN. The sustainable intensification of agriculture. Natural Resources Forum 1997; 21(4) 247-256.

[9] Wivstad M, Dahlin AS, Grant C. Perspectives on nutrient management in arable farming systems. Soil Use and Management 2005; 21 113-121.

[10] Lal R. Soils and food sufficiency: a review. In: Lichtfouse E, Navarrete M, Debaeke Ph, Véronique S, Alberola C. (eds.) Sustainable agriculture. Dordrecht: Springer Science+Business Media BV; 2009. p25-49.

[11] Lal R. Soil degradation as a reason for inadequate human nutrition. Food Security 2009; 1 45-57.

[12] Rabbinge R. The ecological background of food production. In: Crop protection and sustainable agriculture. Ciba Foundation Symposium 1993; 177 222.

[13] Bindraban PS, Stoorvogel JJ, Jansen DM, Vlaming J, Groot JJR. Land quality indicators for sustainable land management: proposed method for yield gap and soil nutrient balance. Agriculture, Ecosystems and Environment 2000; 81 103-112.

[14] Evans L, Fisher R. Yield potential: Its definition, measurement and significance. Crop Science 1999; 39 1544-1551.

[15] Supit I, van Diepen C, de Wit A. Recent changes in the climatic yield potential of various crops in Europe. Agricultural Systems 2010; 103 683-694.

[16] Rabbinge R, Diepen C. Changes in agriculture and land use in Europe. European Journal of Agronomy 2000; 13, 85-100.

[17] Rubio G, Zhu J, Lynch J. A critical test of the prevailing theories of plant response to nutrient availability. American Journal of Botany 2003; 90(1) 143-152.

[18] IFADATA. Available online at: http:// fertilizer.org/HomePage/STATISTICS. (accessed 2012-08-07).

[19] Cakmak I, Kirkby E. Role of potassium nutrition in growth and stress tolerance. Physiologia Plantarum 2008; V 133(4) 692-704.

[20] Marschner H. Mineral nutrition of higher plants. London: Academic Press; 1986.

[21] Grzebisz W, Barłóg P, Feć M. The dynamics of nutrient uptake by sugar beet and its effect on dry matter and sugar yield. Bibliotheka Fragmenta Agronomica 1998; 3(98) 242-249.

[22] Barraclough PB. The growth and activity of winter wheat roots in the field: nutrient uptakes of high-yielding crops. Journal of Agricultural Science Cambridge 1986; 106 45-52.

[23] Cannell RQ. Mitigation of soil constrains to cereal production. In: Gallagher ER. (Ed.). Cereal Production. London: Butterworths; 1984. p193-210.

[24] Barłóg P, Grzebisz W. Growth dynamics and nutrient uptake by winter oilseed rape fertilized with three types of nitrogenous fertilizers and a method of the second nitrogen rate division. Oilseed Crops 2000; XXI(1) 85-96 (in Polish with English summary).

[25] Zerche S, Hecht R. Nitrogen uptake of winter wheat during shoot elongation phase in relation to canopy high and shoot density. Agribiological Research 1999; 52(3/4) 231-250.

[26] Grzebisz W, Pepliński K, Szczepaniak W, Barłóg P, Cyna K. Impact of nitrogen concentration in sugar beet plant parts throughout the growing season on dry matter accumulation patterns. *Journal of Elementology* 2012; 17(3) 389-407.

[27] Orlovius, K. Results on oilseed rape fertilization with potassium, magnesium and sulfur in Germany. In: Grzebisz W. (ed.) Balanced fertilization of oilseed rape – current problems. Poznań: Agricultural University; 2000. p229-239 (in Polish with English summary).

[28] Windt A. Entwicklung des Faserwurzelsystems der Zuckerrübe sowie dessesn Beziehung zum Wachstum der Gesamtpflanze und der Nährstoffaufnahme. Göttingen: Cuvillier Verlag; 1995.

[29] Grzebisz W, Barłóg P, Szczepaniak W, Potarzycki J. Effect of potassium fertilizing system on dynamics of dry matter accumulation by maize. Fertilizers and Fertilization 2010; 40 57-69.

[30] King J, Gay A, Sylvester-Bradley R, Bingham I, Foulkes J, Gregory P, Robinson D. Modelling cereal root systems for water and nitrogen capture: towards an economic optimum. Annals of Botany 2003; 91 383-390.

[31] Jung A, Classeen N. Availability in soil and acquisition by plants as the basis for phosphorus and potassium supply to plants. Zeitschrift für Pflanzenernährung und Bodenkunde 1997; 152 151-157.

[32] Barber SA. Potassium availability at the soil-root interface and factors influencing potassium uptake. In: Munson RD. (ed.) Potassium in Agriculture. Madison: SAS-CSSA-SSSA; 1985. p309-323.

[33] Kuchenbuch R, Classesn N, Junk A. Potassium availability in relation to soil moisture. I. Effect of soil moisture on potassium diffusion, root growth and potassium uptake of onion plants. Plant Soil 1986; 95 221-231.

[34] Nye PB, Tinker PH. Solute movement in the rhizosphere. New York: Oxford University Press; 2000.

[35] Pulford I, Flower H. Environmental chemistry at a glance. Oxford: Blackwell Publishing LtD; 2006.

[36] Claassen N, Steingrobe B. 1999. Mechanistic simulation models for a better understanding of nutrient uptake from soil. In: Mineral nutrition of crops. Zdenko Rengel (Ed.), 327-367.

[37] Szczepaniak W, Grzebisz W, Barłóg P, Cyna K, Pepliński K. Effect of differentiated fertilizing systems on nitrogen accumulation patterns during the growing season – a sugar beet example. Journal of Elementology 2012; 17 (4) (in press).

[38] Gonzalez-Dugo V, Durand JL, Gastal F. Water deficit and nitrogen nutrition of crops. A review. Agronomy for Sustainable Development 2010; 30 529-544.

[39] Marschner H, Kirkby EA, Cakmak J. Effect of mineral nutritional status on shoot-root partitioning of photo-assimilates and cycling of mineral nutrients. Journal of Experimental Botany 1996; 1255-1263.

[40] Agren GI, Franklin O. Root: shoot ratios, optimization and nitrogen productivity. Annals of Botany 2003; 92 795-800.

[41] Debaeke Ph, Aboudrare A. Adaptation of crop management to water-limited environments. European Journal of Agronomy 2004; 21 433-446.

[42] Lawlor DW. Carbon and nitrogen assimilation in relation to yield: mechanisms are the key to understanding production systems. Journal of Experimental Botany 2002; 53(370) 773-787.

[43] Sylvester-Bradley R, Lun G, Foulkes J, Shearman V, Spink J, Ingram J. Management strategies for high yields of cereals and oilseed rape. HGCA Conference: Agronomic Intelligence: the basis for profitable production, 16-17 January 2002, London, Great Britain, 2002.

[44] Subedi K, Ma B. Assessment of some major yield-limiting factors on maize production in a humid temperate environment. Field Crops Research 2009; 110 21-26.

[45] Grzebisz W, Diatta J. Constrains and solutions to maintain soil productivity: A case study from Central Europe. In: Whalen JK.(ed.) Soil fertility Improvement and Integrated Nutrient Management - A Global Perspective. Rijeka: InTech; 2012. p159-182.

[46] Smucker AJM, Aiken RM. Dynamic root responses to water deficits. Soil Science 1992; 154(4) 281-289.

[47] Damm S, Hofmann B, Gransee A, Christen O. Zur wirkung von Kalium auf ausgewählte bodenphysikalishe Eigenschaften and den Wurzeltiegang landwirtschatlicher Kulturpflanzen. Archives of Agronomy Soil Science 2011; 1-19. DOI: 10.1080/03650340.2011.596827.

[48] Gäth, S., Meuser, H., Abitz, C-A., Wessolek. G., Renger/ M. (1989): Determination of potassium delivery to the roots of cereal plants. Zeitschrift für Pflanzenernährung und Bodenkunde; 152 143-149.

[49] D'Andrea K, Otequi M, Cirilo A. Kernel number determination differs among maize hybrids in response to nitrogen. Field Crops Research 2008; 105 228-239.

[50] Subedi K, Ma B. Nitrogen uptake and partitioning in stay-green and leafy maize hybrids. Crops Science 2005; 45 740-747.

[51] Grzebisz W, Musolf R, Szczepaniak W, Barłóg P. Effect of artificially imposed water shortages against a background of potassium supply on dry matter, nitrogen and potassium accumulation by winter wheat. Fertilizers and Fertilization 2009; 34 40-52.

[52] Pommel B, Gallai A, Coque M, Quillere I, Hirel B, Prioul J, Andrieu B, Floriot M. Carbon and nitrogen allocation and grain filling in three maize hybrids differing in senescence. European Journal of Agronomy 2006; 24 200-211.

[53] Soulié Fel, Youssoufi MS, Delenne JY, Voivret C, Saix C. Effect of the crystallization of a solute on the cohesion in granular materials. Powder Technology 2007; 175 43-47.

[54] Gregory PJ, McGovan M. Biscoe PV. Water relations of winter wheat. 2. Soil water relations. The Journal of Agricultural Science, Cambridge 1978; 91 103-116.

[55] Szczepaniak W, Grzebisz W, Barłóg P, Przygocka-Cyna K. Response of maize varieties to increasing nutrient input on the background of natural soil fertility. Fertilizers and Fertilization 2010; 40 127-139.

[56] Monneveux P, Belhassen E. The diversity of drought adaptation in the wide. Plant Growth Regulation 1996; 20 85-92.

[57] Bingham IJ. Soil-root-canopy interactions. Annals of Applied Biology 2001; 138 243-251.

[58] Taylor HM, Klepper B. The role of rooting characteristics in the supply of water to plants. Advances in Agronomy 1978; 30 99-128.

[59] Liu J, Williams J, Zehnder A, Yang H. GEPIC – modeling wheat yield and crop productivity with high resolution on a global scale. Agricultural Systems 2007; 94 478-493.

[60] Agnus JF, van Herwaarden AF. Increasing water use and water use efficiency in dryland wheat. Agronomy Journal 2001; 93 290-298.

[61] Passioura J. Increasing crop productivity when is scarce-from breeding to field management. Agricultural Water Management 2006; 80 176-196.

[62] Grzebisz W, Baer A, Barłóg P, Szczepaniak W, Potarzycki J. Effect of nitrogen and potassium fertilizing systems on maize grain yield. Fertilizers and Fertilization 2010; 40 45-56.

[63] Sheldrick WF, Syers JK, Lingard J. A conceptual model for conducting nutrient audits at national, regional and global scales, Nutrient Cycling in Agroecosystems 2002; 62 61-67.

[64] Wood S, Sebastian K, Scherr SJ. Soil resource condition. Research report: Pilot analysis of global ecosystems: Agroecosystems. Washington: World Resource Institute; 2000. p45-54.

[65] Johnston AE.3. Understanding Potassium and Its Use in Agriculture. Brussels: European Fertilizer Manufacturers Association; 2003. http://www.efma.org/ (accessed 2012-08-15).

[66] Grzebisz W. Crop plants fertilization. Part II. Fertilizers and fertilizing systems. Poznań: PWRiL, 2009, (in Polish).

[67] Sparks DL. Potassium dynamics in soils. Advances in Agronomy 1987; 6 1-63.

[68] Grzebisz W, Gawrońska-Kulesza A. Effect of crop rotation on potassium forms and their bio-availability in log-term trial on a black earth. Part I. Three-course rotation. Polish Journal of Soil Science 1995; XXVIII(2) 111-117.

[69] Benbi DK, Biswas CR. Nutrient budgeting for phosphorus and potassium in a long-term fertilizer trial. Nutrient Cycling in Agroecosystems 1999; 54 125-132.

[70] Pagel H. 1972. Vergleichende Untersuchungen über Gehalt an austauschbarem und nachlieferbaren Kalium in wichtigen Böden der ariden und humiden Tropen. Beit. Tropische und sub-tropische Landwirtschaft und Tropenveterinärmedizin, Leipzig: KMU; 1972; 10 35-51.

[71] Fotyma, M. (2007) Content of potassium in different forms in the soils of Poland. Polish Journal of Soil Science 2007; XL(1) 19-32.

[72] Manning DAC. Mineral sources of potassium for plant nutrition. A review. Agronomy for Sustainable Development 2010; 30 281-294.

[73] Robert M. 1993. K-fluxes in soils in relation to parent material and pedogenesis in tropical, temperate and arid climates. In: Potassium in Ecosystems. Proceedings of

the 23rd Colloquium of the International Potash Institute, 12-16 October 1992, Praha, the Czech Republic, International Potash Institute, Basel, p25-44.

[74] Jakovljević MD, Kostić NM, Antić-Mladenović SB. The availability of base elements (Ca, Mg, Ma, K) in some important soils types in Serbia. Proceedings for natural Sciences 2003, Matica Srpska Novi Sad; 104 11-23

[75] Sheng XF. Growth promotion and increased potassium uptake of cotton and rape by a potassium releasing strain of Bacillus edaphicus. Soil Biology and Biochemistry 2005; 37 1918-1922.

[76] Basak BB, Biswas DR. Influence of potassium solubilizing microorganism (Bacillus mucilaginosus) and waste mica on potassium uptake dynamics by sudan grass (Sorghum vulgare Pers.) grown under two alfisols. Plant Soil 2009; 317 (1/2) 235-255.

[77] Wang JG, Zhang FS, Zhang XL, Cao YP. Release of potassium from K-bearing minerals: effect of plant roots under P deficiency. Nutrient Cycling in Agroecosystems 2000; 56 45-52.

[78] Frąckowiak-Pawlak K. On bio-fertilizer effects on soil and plant properties. In: Report 2006/2007 Poznan University of Life Sciences, 2008.

[79] Bennet AJ, Bending GD, Chandler D, Hilton S, Mills P. Meeting the demand for crop production: the challenge of yield decline in crops grown in short rotations. Biological Reviews 2011. DOI: 10.1111/j.1469-185X.2011.00184.x

[80] Berzsenyi Z, Győrffy B, Lap DQ. Effect of crop rotation and fertilization on maize and wheat yields and yield stability in a long-term experiment. European Journal of Agronomy 2000; 13 225-244.

[81] Karlen DL, Varvel GE, Bullock DG, Cruse RM. Crop rotations for the 21st century. Advances in Agronomy 1994; 53 1-45.

[82] Castellazzi MS, Wood GA, Burgess PJ, Morris J, Conrad KF, Perry JN. A systematic representation of crop rotations. Agricultural Systems 2008; 97 26-33.

[83] Struik P, Bonciarelli F. Resource use at the cropping system level. European Journal of Agronomy 1997; 7 133-143.

[84] Weaver JE. Root development of field crops. New York: McGraw-Hill BC Inc. 1926.

[85] Milford GFJ, Armstrong MJ, Jarvis PJ, Houghton BJ, Bellett-Travers DM, Jones J, Leigh RA (2000): Effect of potassium fertilizer on the yield, quality and potassium off-take of sugar beet crops grown on soil of different potassium status. The Journal Agricultural Science, Cambridge 2000; 135 1-10.

[86] Wojciechowski A, Szczepaniak W, Grzebisz W. Effect of potassium fertilization on yields and technological quality of sugar beet part I. Yields of roots and sugar. Biuletyn IHAR 2002; 222 57-64 (in Polish with English summary).

[87] Heckman JR, Kamprath EJ. Potassium accumulation and corn yield related to potassium fertilizer rate and placement. Soil Science Society of American Journal 1992; 56 141-148.

[88] Van Noordwijk M, Brouwer G. Review of quantitative root length data in agriculture. In: McMichael BL, Persson H. (eds.) Plant roots and their environment. Amsterdam: Elsevier Science Publishers; 1991. p515-525.

[89] Van Noordwijk M, de Willigen P. Roots, plant production and nutrient use efficiency. PhD thesis. Institute of Soil Fertility, Haren (Groningen); 1987.

[90] Kerschberger M, Richter D. Neue Versorgungsstufen (VST) für den pflanzenverfügbaren K-Gehalt (DL-Methode) auf Ackerböden. Richtlinien der Düngung 1987; 11, 14-18.

[91] Janssen BH, de Willigen P. Ideal and saturated soil fertility as bench marks in nutrient management. 1. Outline of the framework. Agriculture, Ecosystems and Environment 2006; 116 132-146.

[92] Wojciechowski A, Szczepaniak W, Grzebisz W. Effect of potassium fertilization on yields and technological quality of sugar beet. Part III. Potassium uptake. Biuletyn IHAR 2002; 222 71-76.

Chapter 9

Alternative Fertilizer Utilizing Methods for Sustaining Low Input Agriculture

Monrawee Fukuda, Fujio Nagumo,
Satoshi Nakamura and Satoshi Tobita

Additional information is available at the end of the chapter

1. Introduction

1.1. Declining soil fertility in low input agriculture

Improvement of soil fertility and plant nutrition to sustain adequate yield of crop is essential since soil degradation has been identified as a major constraint and a root cause of declining crop productivity in many developing countries *e.g.* Sub-Saharan Africa (SSA). Sanchez [1] reported very high rate of annual depletion for 22 kg nitrogen (N), 2.5 kg phosphorus (P), and 15 kg potassium (K) per hectare of cultivated land or an annual loss equivalent to 4 billion U.S. dollar in fertilizer in 37 African countries over three decades. Due to large quantities of nutrients are removed from soil through crop harvest without sufficient supply of fertilizers and manure causing low input agriculture has been unfortunately implemented by farmers and the consequences of low crop productivity would increase food insecurity. In many regions, local farmers lack of sufficient fertilizer, money for purchase, access to the credit, and transportation resulting to low in fertilizer input and a gradual decrease of soil fertility [2]

1.2. Limitation on replenishing soil fertility and increasing crop yield

Numbers of strategies have been used to restore soil fertility including traditional application of inorganic fertilizers or use of organic fertilizing materials such as plant residues (*i.e.* rice straw and husk), green manure, and animal manure [3]. Uses of crop management system such as cover crops, legumes, mulching, fallow, and agroforestry are well documented [4]. Moreover, adoption of high yielding and genetically improved crop varieties is a good option for increasing yield productivity.

Amongst ways of soil fertilization, increasing use and continuous application of inorganic fertilizers seemed to be limited because fertilizers in Africa are 2 – 6 times more expensive than that of in Europe, North America, and Asia [1]. Applying plant residue or organic biomass to soil has influenced on soil nutrients, soil physical condition, soil biological activity, and crop performance. However, applying these organic fertilizing materials such as rice straw and husk, green manure or organic biomass (*i.e.* leaf biomass) to soil are not attractive to farmers compared to straw burning due to short term effects of organic materials on crop yield are often small. Cutting and carrying biomass to the field also require high labor and cost. On other hands, crop residues have high economic value and have been used as livestock feed and fuel so leaving crop residues in the field is seldom. Even though, incorporation of rice straw which is abundant and widely spread in the rice field can return and reserve most of nutrients to soil particularly N, P, K, S, and Si in long term [5]. Tobita et al. [6] and Issaka et al. [3] reported that adding rice straw to rice system could gain approximately 20 percent of N and P, and most K relative to the needs of applied chemical fertilizers in the Northern region of Ghana where rice cultivation is the most prominent.

Crop management such as tree fallow system is not attractive for farmer because they prefer better land use alternative owing to population pressure particularly in the humid and tropical regions. Besides, improved fallows have not been proved yet on their benefits in semiarid tropics of Africa. The potential of fallow system on shallow and poorly drained soil is poor [1]. Growing leguminous plants as fallows before cropping season or intercropping with crop is effective crop management to accumulate N for consecutive crops. However, it should be noted that effects of plant residues on soils and crops depend on the quality (*i.e.* carbon/nitrogen ratio, lignin, and polyphenol contents) and the decomposition rates of residues which in turn control the nutrient release rates. Tian et al. [7] found that the contribution of low quality plant residues as mulching on maize grain yield and protein concentration was lowest in comparison to intermediate or high quality residues on Oxic Paleustalf soil in Nigeria.

Animal manures from poultry, pig, cow, goat, and sheep contain all the major nutrients. These manures are very good materials for improving soil fertility and crop productivity [8,3]. Tobita et al. [6] reported that if only 20 percent of total livestock organic resource estimated in Ghana was utilized, so it could replace the requirement for chemical fertilizer in rice cultivation system entire the Northern region. However, gathering bulky dung of livestocks or excreta (dung and urine) from grazing livestock was difficult particularly in rural area where these manures are not sold and scarce [9]. Unlikely, poultry manure may be valid in urban center where intensive production of poultry has being implemented. In present, poultry manure is on high demand but its quantity is not enough for farmer's need resulting farmers have to pay in advance before manure will be delivered to the field [3,10].

1.3. Alternative P fertilizer utilizing methods

Many soils in sub-humid and humid tropics including SSA have very low levels of natural P, thus P fertilization is essential for maintaining desired level of crop yield. Buresh et al. [2] indicates that input of P fertilizers is required to replenish P stock in highly P deficient soils rather than only dependence on P cycling through organic-based system. However, P fertilizer management became more difficult because the only natural P source *i.e.* phosphate rock (PR) for manufacturing P chemical fertilizers is non-renewable and finite resource. Though, ground PR can be directly used as P fertilizer but it slowly releases P in acid soils resulting in gradual build up P in numbers of cropping season.

Phosphate rock, a natural form of mineral apatite contains not readily phosphate content for plant. Phosphate rock must be treated to convert phosphate to water soluble of plant available forms [11]. Major solid water soluble P fertilizers are single superphosphate (SSP), triple superphosphate (TSP), monoammonium phosphate (MAP), and diammonium phosphate (DAP). In fully acidulated commercial grade P fertilizers, SSP is made by adding sulfuric acid to PR. Triple superphosphate containing about 2 times of P concentration as SSP, is made by adding phosphoric acid to PR. Ammonium phosphate fertilizers are produced by passing amnonia through phosphoric acid. The compounds of P within water soluble fraction are mainly in the forms of monocalcium phosphate or MCP [$Ca(H_2PO_4)_2.2H_2O$] in SSP and TSP, $NH_4H_2PO_4$ in MAP, and $(NH_4)_2HPO_4$ in DAP in which over 90% of total P concentrations are water soluble P [12,13]. Diammonium phosphate, MCP, and TSP can be accounted as a half of phosphate-based fertilizer applications worldwide [11].

Continuous mining PR and increasing use of P chemical fertilizers might not be responsible management of resource use. More efficient uses of P fertilizers in agriculture have been paid intention. Phosphorus efficient plants are recently developed through plant breeding or genetic modification, but more P efficient plants which are modified root growth and architecture, manipulated root exudates, or managed plant-microbial association such arbuscular mycorrhizal fungi and microbial inoculants are not common, and still have less potential trade-off [14]. So far, fertilizer management still has a significant contribution to overall agricultural crop production, household's farming system, farmers, and the rural poor [4]. In addition, the potential of genetically improved crops cannot be achieved when soils are depleted of nutrients. Sanchez [1] stated that improved crop varieties have responsible for only 28% yield increases in Africa, but 66-88% in Asia, Latin America, and the Middle East when rates of adoption for new improved varieties have been similar during the last four decades. Therefore, it is necessity to find alternative fertilizer utilizing methods to increase and sustain crop yield in low input agriculture and these methods should be affordable for local farmer.

2. Materials and methods

This work gathered information from published papers (secondary data) focusing on the utilization of small quantity of fertilizer to boost crop productivity in wide-range of climates and soil conditions. The effective methods have been revealed including 1) fertilizer microdose application, 2) addition of small amount of fertilizer to the seed by coating, 3) increase of nutrient concentration in seedling by soaking in, or dipping seedling in the nutrient slurry. Moreover, two experiments were conducted to investigate the effects of fertilizer seed coating and fertilizer seedling soaking on the early growth of rice (*Oryza sativa* cv. IR74) grown on acidic P deficit soil.

2.1. Fertilizer seed coating method

Treatments were triplicated and comprised of 1) control_uncoated; 2) control_oil; 3) Burkina Faso phosphate rock (BPR); 4) Potassium dihydrogenphosphate (KH_2PO_4) and 5) NPK (14-14-14). The 15 seeds of IR74 rice were coated by 2 levels of ground fertilizer (18 or 36 mg) using vegetable oil as adhesive material. By this method, ground fertilizers were mixed with seeds at approximately 1.2 and 2.4 mg per seed. Coated seeds were sowing directly into moistened soil. Soil used in this experiment was collected from Tropical Agriculture Research Front (TARF). Massive amount of soil was collected, air dried, and sieved to 2 mm. Five hundred gram of air dry soil was weighed into bag. Soil properties were pH_{H2O}, 4.83; EC, 6.03 mS m^{-1}, and Bray 1-P, 1.66 mg kg^{-1}. The basal nutrients were mixed to each soil (mg kg^{-1} soil); $Na_2MoO_4.2H_2O$; 0.36; H_3BO_3, 0.71; $CuSO_4.5H_2O$, 5; $ZnSO_4.7H_2O$, 10; $MnSO_4.H_2O$, 15; $MgSO_4.7H_2O$, 21; $CaCl_2.2H_2O$, 71; K_2SO_4, 142; NH_4NO_3, 200, respectively. Percentage of plant emergence was monitored at 5, 10, and 15 day after sowing (DAS), and then thinned to 4 seedlings. Watering was done daily. Plant height, tiller number, and leaf age were measured at 20 and 40 DAS. Two rice plants were sampled for shoots and roots from each treatment at 20 and 40 DAS, oven-dried to obtain dry matter, and ground prior to further chemical analysis. Total P in plant organs was determined after dry-ashing procedure.

2.2. Fertilizer seedling soaking

Treatments were triplicated and comprised of 1) control (+P soil); added P soil [adding $Ca(H_2PO_4)_2$; 331.4 mg] with non-fertilizer soaked rice (conventional method by farmers); 2) control (-P soil); non P added soil with non-fertilizer soaked rice, and other treatments were conducted on non added P soil with soaked rice by 3) Potassium dihydrogenphosphate (KH_2PO_4) solution; 4) NPK (14-14-14) solution. Five seedlings of 6-7 leaf age rice were soaked in 1 % and 5% (w/v) of each fertilizer solution with 2 soaking periods (30 min or 60 min). Freshly soaked seedlings were transplanted directly into flooded soil. Rice seedlings were thinned to 3 seedlings at 7 days after transplanting (DAT). Rice seedlings used in this experiment were grown on fully fertilized soil for 3 weeks before soaking and transplanting into pots. Three kilogram of 2 mm sieved soil was weighed into 1/5,000 are-pot. The basal nutrients were mixed to soil at the same rate of previous experiment described above. Water was added daily to maintain submerged/flooded condition. Tiller number and leaf age were monitored at 20 and 40 DAT. Plant height was measured at 20, 40, and 75 DAT, and then a rice plant was harvested for shoot and root at 20 and 40, and 75 DAT, then oven-dried to obtain dry matter, and ground prior to further analysis for total P concentration after dry ashing procedure.

2.3. Data and statistical analysis

The JMP 9.0.0 (SAS Institute Inc., USA) was used to perform ANOVA and compare the means by the Tukey Kramer HSD for plant growth (height, tiller number, leaf age), shoot and root DM, P concentration, P uptake.

3. Results and discussion

3.1. Fertilizer microdosing

Fertilizer microdosing or known as microdose fertilizer application or point application is an application method of small, affordable quantity of fertilizer with the seed at planting time or as top dressing 3-4 weeks after emergence [15]. This application method has been developed by the International Crops Research Institute for the Semi-Arid Tropics (ICRISAT) and its partners to improve inorganic fertilizer use of farmers in Sahel region, Africa. This method is documented to enhance fertilizer use efficiency, high probability of yield response, crop productivity with favorable fertilizer per grain price ratio rather than spreading fertilizer over the field, root systems, and soil water capture [15,16].

Microdosing soil with fertilizer uses about one-twentieth of the amount of fertilizer used on corn, and one-tenth of the amount used on wheat in America. Particularly, this pity amount often doubles crop yields on African soils due to they are starved of macronutrients such as N, P, and K. Small doses of fertilizers, about a full bottle cap or a three-finger pinch per a hole of planting are required and this amount equals to 6 gram of fertilizer or about 67 pound of fertilizer for every 2.5 acres or 30 kg fertilizer per hectare. Farmers just prepare small holes before the rain starts when soils are still hard. Later, fertilizers and seeds shall be put in the hole when the rain begins and the soils provide enough moist condition, encouraging root growth [15].

Successful works have been showed through Tabo et al. [17] who reported the yields of sorghum and millet were increased from 44 to 120% after adoption of fertilizer microdosing in harsh semi-arid climate of Mali, Burkina Faso, and Niger, the western Africa where soils were sandy and low in fertility with 500-800 mm annual rainfall. Farmers, themselves, selected plant varieties and types of fertilizer which were differed among countries and availability of fertilizer on the local markets. Rates of fertilizer micro-dose per hill of planting were 4 g of NPK (15-25-15) in Burkina Faso, 4 g of NPK (17-17-17) in Mali, and 6 g of NPK (15-15-15), 2 g DAP (18-46-0), and 2 g DAP + 1 g Urea (46-0-0) in Niger.

Bagayoko et al. [16] compared the effectiveness of fertilizer microdosing among no fertilizer microdosing check (farmer's practice), blank (zero fertilizer application), microdosing only, microdosing plus 20 kg P_2O_5 ha^{-1} and 30 kg N ha^{-1} in wide range of climates and soils in Burkina Faso, Mali, and Niger. It was found that microdose fertilizer application increased yields of grain and stover of pearl millet across wide range of climates and soils in Burkina Faso, Mali, and Niger. Additional supply of 20 kg P_2O_5 ha^{-1} and 30 kg N ha^{-1} had much increased grain and stover yields of pearl millet. Fertilizer microdose rates were 4 g of NPK (15-15-15) or equivalent to 62.5 kg ha^{-1} in Burkina Faso, 2 g of DAP (18-46-0) or equivalent to 33.2 kg ha^{-1} in Mali, and 4 g of NPK (15-15-15) or equivalent to 62.5 kg ha^{-1} in Niger. Nutrient sources of N as urea (46-0-0) and P as 0-46-0 were additionally supplied.

Hayashi et al. [18] demonstrated that millet farmers could delay inorganic fertilizer application or timing of using the micro-dosing technology from 10 to 60 days after sowing without the reduction of profits and their economic returns relative to the non-fertilizer applied treatment. Fertilizer microdose rate was applied 6 g of NPK (15-15-15) per millet hill or 60 kg NPK per hectare for an on-station trial, and 2 g of DAP (18-46-0) per millet hill or 7.24 kg of DAP per hectare for an on-farm field trial. The results stressed that local farmers had more options of fertilizer utilization timing. Delayed fertilizer microdosing still increased millet production and helped farmers who were not able to supply fertilizer at sowing or suffered from shortage of labors and fertilizers. The outcome of this work showed more flexibility in managing money and labor resources for purchasing fertilizer. Microdosing technique was more advantage than other methods to increase crop productivity for subsistence farmer in harsh Sahel region.

3.2. Fertilizer seed coating

According to the survey on inorganic fertilizer application practices by farmers in Fakara, Niger, West Africa [18], it showed that every farmer's household applied fertilizer by mixing fertilizer with seed before planting. Although, mixing rate of fertilizer and seed was very low (fertilizer/seed = 0.2) or equivalent to 0.9-1.8 kg of fertilizer per hectare indicating that farmers have attempted to mix very little fertilizer which they could afford with seeds in order to plant as vast an area as possible. Farmers were aware of fertilizing soil but they were not able to purchase sufficient amount of fertilizer due to some credit and financial problems. Amount of applied fertilizer at 0.9-1.8 kg by farmer's practicing was less than that of recommended level at 9 kg P_2O_5 per hectare by microdosing method which was essential to obtain the optimal improvement of millet production. Therefore, farmers could not achieve desired levels of crop, but some residual effects on P in soil could be expected after this kind of practices. From this view point, it should be noted that farmers really lacked of adequate amount of fertilizer to be used although an effective fertilizer utilizing method such as microdosing has been introduced, but it still consumes quantity of fertilizer and labor. Therefore, another fertilizer utilizing method should be considered in order to reduce much quantity of fertilizer and even labor requirement.

Up-to date, a method such fertilizer seed coating with use of very pity quantity of fertilizer has been interested as an alternative method [15] This method applies ground fertilizer on seed using sticky adhesive materials to firmly attach fertilizer on seed. Fertilizer seed coating may have advantage over mixing fertilizer with seed due to lower labor requirement and high concentration of seed nutrients may be easily raised after firmly coating seed. The release of nutrients from fertilizer coated seed is expected to be much closure to plant root rather than mixing fertilizer and seed before planting. Besides, high concentration of seed nutrients are important for plant establishment in soil which low in nutrient availability, as a massive root system is needed before soil can supply sufficient nutrients to meet the needs of plant [19].

Ros et al. [19] pointed out that the effective of P fertilizer on early plant growth was enhanced by coating rice seed (*Oryza sativa* cv. IR66) with various P fertilizers. Inorganic P fertilizers used for seed coating included single superphosphate (SSP), phosphate rock (PR), monoammonium phosphate (MAP), and potassium phosphate (KH_2PO_4; PP). The rates of applied P fertilizer in mg P per seed were 3.8 coating-SSP; 1.2 coating-PR; 3.4 coating-MAP;

3.7 coating-PP and methyl cellulose glue at the rate of 5% (w/v) was used as adhesive material. The results revealed that coating rice seed increased shoot dry matter (DM) but decreased root DM at 20 days after sowing (DAS) and the effect of coating persisted to 40 (DAS), root length and DM also increased, moreover shoot DM increased 400-870% at this stage. Coating rice seed with PR was more promising for stimulating early growth of rice on low P soils. Coating rice seed by 1.2 mg PR per seed or 0.5 kg PR per kg of seed was not harsh to seedling emergence, but increased a fourfold higher shoot and root growth of rice.

From our works [10] attempted to coat rice seed (*Oryza sativa* cv. IR74) by 1.2 or 2.4 mg per seed of ground fertilizers: Burkina Faso phosphate rock (BPR), Potassium Dihydrogenphosphate (KH$_2$PO$_4$), NPK (14-14-14) before direct sowing. The results revealed that coating rice seed by powdered KH$_2$PO$_4$ for 1.2 or 2.4 mg per seed using vegetable oil as adhesive material could increase plant DM to 174 and 215% (Table 1, Figure 1, and Photo 1), height to 142 and 131%, and leaf age to 118 to 120% (Table 2) at 40 DAS, shoot P concentration to 172 and 226%, P uptake to 136 and 160% at 20 DAS (Table 1), and shoot P concentration to 196 and 168%, and P uptake to 336 and 359% compared to the control (without coating) at 40 DAS (Table 1), respectively. Moreover, plant root DM and P uptake increased to 164 and 199% with 2.4 mg KH$_2$PO$_4$ compared to the control (without coating) at 40 DAS, respectively (Table 3).

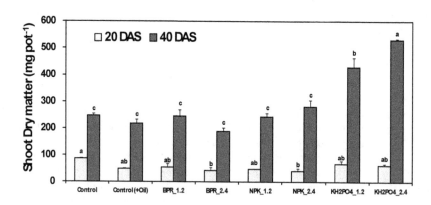

Figure 1. Shoot dry matter of rice after fertilizer seed coating

Treatment	Dry matter (mg)				P concentration (mg kg^{-1} DM)				P uptake (mg pot^{-1})			
	Days after sowing (DAS)											
	20		40		20		40		20		40	
	Mean	SE	Mean	SE	Mean	SE	Mean	SE	Mean	SE	Mean	SE
Control	86.8 ±	2.7 a	248 ±	7 c	814 ±	28 b	451 ±	51 b	0.07 ±	0.0 a	0.11 ±	0.0 c
Control (+Oil)	49.8 ±	3.0 ab	218 ±	15 c	1354 ±	211 ab	601 ±	128 ab	0.07 ±	0.0 a	0.13 ±	0.0 bc
BPR_1.2	55.5 ±	12.9 ab	245 ±	24 c	1319 ±	160 ab	601 ±	67 ab	0.07 ±	0.0 a	0.15 ±	0.0 bc
BPR_2.4	42.5 ±	10.9 b	190 ±	13 c	1587 ±	269 ab	733 ±	113 ab	0.06 ±	0.0 a	0.14 ±	0.0 bc
NPK_1.2	47.4 ±	2.0 ab	243 ±	14 c	1632 ±	58 ab	622 ±	27 ab	0.08 ±	0.0 a	0.15 ±	0.0 bc
NPK_2.4	40.3 ±	9.9 b	281 ±	25 c	1587 ±	61 ab	741 ±	65 ab	0.06 ±	0.0 a	0.21 ±	0.0 b
KH2PO4_1.2	67.2 ±	12.9 ab	431 ±	35 b	1400 ±	59 ab	886 ±	86 a	0.10 ±	0.0 a	0.38 ±	0.0 a
KH2PO4_2.4	63.0 ±	4.6 ab	532 ±	5 a	1840 ±	353 a	756 ±	3 ab	0.11 ±	0.0 a	0.40 ±	0.0 a

Different letters showed significant differences at 0.05% levels by the Tukey Kramer HSD ($n=3$)

Table 1. Shoot dry matter, P concentration, and P uptake of rice plant at 20 and 40 days after sowing as affected by fertilizer seed coating

Treatment	Height (cm)				Leaf age			
	Days after sowing (DAS)							
	20		40		20		40	
	Mean	SE	Mean	SE	Mean	SE	Mean	SE
Control	20.3 ± 1.1 a		27.7 ± 0.6 b		5.3 ± 0.3 a		7.1 ± 0.1 c	
Control (+Oil)	16.3 ± 1.3 a		30.8 ± 1.1 ab		3.4 ± 0.0 a		6.9 ± 0.1 c	
BPR_1.2	22.9 ± 2.6 a		28.7 ± 0.8 b		4.8 ± 0.1 a		7.0 ± 0.0 c	
BPR_2.4	20.9 ± 2.4 a		28.2 ± 5.5 b		4.8 ± 0.2 a		7.3 ± 0.2 bc	
NPK_1.2	23.3 ± 2.0 a		31.7 ± 2.5 ab		5.0 ± 0.0 a		7.9 ± 0.1 abc	
NPK_2.4	21.7 ± 1.7 a		35.0 ± 3.0 ab		4.7 ± 0.3 a		8.0 ± 0.1 ab	
KH2PO4_1.2	25.3 ± 2.0 a		39.3 ± 3.5 a		5.2 ± 0.2 a		8.4 ± 0.2 a	
KH2PO4_2.4	25.4 ± 5.0 a		36.2 ± 3.3 ab		5.1 ± 0.1 a		8.5 ± 0.4 a	

Different letters showed significant differences at 0.05% levels by the Tukey Kramer HSD ($n=3$)

Table 2. Height and leaf age of rice plant at 20 and 40 days after sowing as affected by fertilizer seed coating

20 DAS   **40 DAS**

Photo 1. The growth of rice plant after fertilizer seed coating

Treatment	Dry matter (mg)				P concentration (mg kg⁻¹ DM)				P uptake (mg pot⁻¹)			
	Days after sowing (DAS)											
	20		40		20		40		20		40	
	Mean	SE	Mean	SE	Mean	SE	Mean	SE	Mean	SE	Mean	SE
Control	14.2 ± 1.5 a		176 ± 14 b		790 ± 74 a		604 ± 14 a		0.011 ± 0.0 a		0.11 ± 0.0 bc	
Control (+Oil)	7.0 ± 1.1 b		129 ± 0.1 b		1032 ± 52 a		714 ± 76 a		0.007 ± 0.0 a		0.09 ± 0.0 c	
BPR_1.2	6.2 ± 0.9 b		146 ± 13 b		1059 ± 133 a		753 ± 70 a		0.006 ± 0.0 a		0.11 ± 0.0 bc	
BPR_2.4	6.0 ± 1.4 b		148 ± 26 b		1377 ± 921 a		808 ± 104 a		0.014 ± 0.0 a		0.12 ± 0.0 bc	
NPK_1.2	6.2 ± 0.5 b		148 ± 23 b		1093 ± 104 a		786 ± 70 a		0.007 ± 0.0 a		0.12 ± 0.0 bc	
NPK_2.4	7.1 ± 1.6 b		133 ± 23 b		1441 ± 248 a		807 ± 60 a		0.009 ± 0.0 a		0.11 ± 0.0 bc	
KH2PO4_1.2	9.5 ± 1.3 ab		229 ± 39 ab		1035 ± 89 a		856 ± 63 a		0.010 ± 0.0 a		0.19 ± 0.0 ab	
KH2PO4_2.4	7.8 ± 0.9 b		288 ± 11 a		1259 ± 218 a		739 ± 28 a		0.009 ± 0.0 a		0.21 ± 0.0 a	

Different letters showed significant differences at 0.05% levels by the Tukey Kramer HSD ($n=3$)

Table 3. Root dry matter, P concentration, and P uptake of rice plant at 20 and 40 days after sowing as affected by fertilizer seed coating

Coating dry rice seed by ground/powdered KH_2PO_4 with seed with small volume of vegetable oil extended the growth of seedling up to 40 DAS in soil where none of P fertilizer was applied. Delayed plant emergence could be found at 5 DAS but this problem was overcome after 10 days. The growth of plant was enhanced after 20 days compared to un-coated plant. Coating rice seed by 2.4 mg powdered KH_2PO_4 per seed or 92 g KH_2PO_4 per kg of seed (averaged seed weigh = 26.08 mg) was expected to be low cost and easily handled.

Use of BPR or NPK, some procedures such as pre-germination or dormancy break of rice seed might be required to increase water imbibition, and subsequent emergence, and root growth prior to nutrient released from these P fertilizers could supply adequate amount of P to plant without damaging effect on the growth.

3.3. Fertilizer seedling dipping/soaking

Lowland rice soils in the tropics are P deficient and the management of P fertility in soils depends on P source, timing, and application method. More effective methods for P application are surface broadcasting or incorporation of fertilizer before planting rather than deep placement of P at 10 or 20 cm depth in planting hill or between planting rows. While, the best timing of applying P fertilizer for rice is at transplanting with total dose of a basal P because plant requires more P at early growth stage. Sufficient P supply may increase better root development and tillering. However P fertilizer application may also be delayed, but it should be before the vigorous stage of tillering. Split application method of P is less effective and not necessary due to P mobility from old leaves to new ones. In contrast, applying P fertilizer 2 weeks before panicle initiation of rice plant is as effective as that applied at transplanting. It is considered that 8-20 % of fertilized P to soil is recovery by rice and remaining 80-90% of applied P can benefit to succeeding crops [20].

Methods of P fertilization with use of small quantity such as fertilizer seedling soaking/dipping has drawn attention in some countries [21]. Lu et al. [21] stated that dipping rice seedling in phosphate fertilizer was a traditional method in China. Farmers generally applied P by mixing with fertile soil or compost in a portion of 1:1 or 1:5 and water to make a paste or slurry. Rice seedlings were dipped into this slurry before transplanting however it was necessary to avoid damage of root during dipping. Another work by Katyal [22] cited by [21] showed dipping seedling roots might provide 40-60% saving on P fertilizer for maintaining the same level of yield. Ling [23] cited by [21] showed that P fertilizer recovery has been markedly increased following dipping rice seedling roots by using ^{32}P experiment. The effects of dipping/ soaking seedling in P fertilizer may be attributed to a direct contact of rice roots with P fertilizer resulting in a greater gradient of P concentration was established and would facilitate the diffusion of P to the roots [21]. Since, rice plant during early growth stages required more P, but available P from soil could not meet the needs of plant at this stage. Therefore, enhancing plant's early growth stage by fertilizer seedling soaking/dipping would increase root development and tillering and in turn increased rice grain yield, particularly in P deficit soils [20]. Besides, De Datta et al. [20] reported that dipping rice seedling root in a P-soil slurry reduced fertilizer requirement by 50%. Katyal [22] cited by [20] indicated P fertilizer utilization was reduced to 50% without decreasing yield with this dipping seedling method. Therefore, application of P to root in form of a slurry before transplanting was an economical method.

From our work [10] showed the soaking rice seedling (*Oryza sativa* cv. IR74) in P fertilizer solution before transplanting increased the growth of rice grown on acidic P deficit soil up to 75 DAT. The procedure of fertilizer seedling soaking has been showed in Photo 2.

Under non-P fertilized soil, soaking rice seedling with 5% KH_2PO_4 solution before transplanting for 30 and 60 min increased shoot DM to 246 and 235%, shoot P concentration to 159 and 141%, root P concentration to 155 and 135 %, leaf age to 117 and 119%, and tiller number to 300 and 433 % at 20 days after transplanting (DAT), respectively (Table 4, 5, 8, 11, 12 and Figure 2). At 40 DAT, shoot DM and P uptake, root concentration and P uptake, leaf age, and tiller number increased to 265, 277, 456, 471, and 115, and 375%, respectively with

5% KH$_2$PO$_4$ for 60 min (Table 4, 6, 8, 9, 11, and 12). At 75 DAT, shoot DM was increased by soaking with 1 and 5 % KH$_2$PO$_4$ to 141 and 331% for 30 min, and 167 and 181 % for 60 min compared to the control (-P) soil, respectively (Figure 2, Table 4). Root DM was increased by soaking with 5% KH$_2$PO$_4$ to 299 and 138% for 30 and 60 min soaking, respectively. By soaking with 1% KH$_2$PO$_4$ for 30 min increased root DM to 115% (Figure 2, Table 7).

Photo 2. Procedure of fertilizer seedling soaking

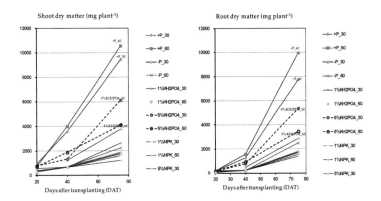

Figure 2. Dry matter of rice's shoot and root after fertilizer seedling soaking

Shoot P uptake was increased after soaking by 1 and 5 % KH_2PO_4 to 147 and 292 % for 30 min, and 204 and 100 % for 60 min, respectively (Table 6). Root P uptake was increased by soaking by 1 and 5% KH_2PO_4 to 130 and 324 % for 30 min, and 131 and 159 % for 60 min, respectively (Table 9). This revealed the seedling soaked by 5% KH_2PO_4 for 30 min has progressive increased shoot and root DM and P uptake from early growth stage to 75 DAT. In contrast, 5 % NPK severe damaged seedling and caused to death of seedling. While, 1 % NPK had no effect on the growth compared to the control. This study concluded that soaking rice seedling with 5% KH_2PO_4 solution before transplanting for 30 min improved the growth of rice up to 75 DAT on lowland acidic P deficit soil without P fertilization. However, it should be noted that the fertilization soil with fertilizer was required to maintain desired level of rice production. Sufficient supplied nutrients support entire crop's life cycle.

Treatment	Shoot Dry matter (mg)											
	20 DAT				40 DAT				75 DAT			
	30 min		60 min		30 min		60 min		30 min		60 min	
	Mean	SE	Mean	SE	Mean	SE	Mean	SE	Mean	SE	Mean	SE
Control (+P soil)	955 ± 273 a		737 ± 45 abc		3583 ± 973 a		3997 ± 166 a		9442 ± 1446 a		10552 ± 2218 a	
Control (-P soil)	322 ± 11 c		315 ± 17 c		705 ± 132 b		709 ± 80 b		1854 ± 472 bc		2252 ± 298 bc	
1% KH_2PO_4	573 ± 26 abc		507 ± 49 bc		666 ± 36 b		1241 ± 156 b		2622 ± 603 bc		3768 ± 661 bc	
5% KH_2PO_4	792 ± 42 ab		741 ± 43 abc		1337 ± 163 b		1880 ± 167 b		6142 ± 648 ab		4108 ± 230 bc	
1% NPK	389 ± 25 bc		329 ± 18 c		674 ± 181 b		683 ± 63 b		1602 ± 556 bc		1225 ± 194 c	
5% NPK	306 ± 21 c		- ±		686 ± 44 b		- ±		1759 ± 72 bc		±	

Different letters in the same day after incubation (DAT) showed significant differences at 0.05% levels by the Tukey-Kramer (n=3)

Table 4. Shoot dry matter of rice plant at 20, 40, and 75 days after transplanting as affected by various fertilizers and timing of seedling soaking

Treatment	Shoot P concentration (mg kg⁻¹ DM)											
	20 DAT				40 DAT				75 DAT			
	30 min		60 min		30 min		60 min		30 min		60 min	
	Mean	SE	Mean	SE	Mean	SE	Mean	SE	Mean	SE	Mean	SE
Control (+P soil)	1665 ± 64 a		1423 ± 93 a		755 ± 23 a		631 ± 25 ab		1411 ± 45 a		1169 ± 92 ab	
Control (-P soil)	642 ± 66 d		764 ± 24 cd		600 ± 15 ab		489 ± 26 b		757 ± 71 cd		751 ± 26 cd	
1% KH₂PO₄	765 ± 35 cd		772 ± 43 bc		508 ± 34 b		506 ± 16 b		806 ± 30 cd		903 ± 34 bcd	
5% KH₂PO₄	1019 ± 51 b		1078 ± 62 b		546 ± 10 b		508 ± 28 b		1013 ± 42 bc		836 ± 20 cd	
1% NPK	656 ± 54 d		845 ± 50 bcd		539 ± 29 b		593 ± 37 b		721 ± 107 d		686 ± 46 d	
5% NPK	883 ± 81 bcd		-		569 ± 62 b		-		756 ± 6 cd			

Different letters in the same day after transplanting (DAT) showed significant differences at 0.05% levels by the Tukey-Kramer (n=3)

Table 5. Shoot P concentration of rice plant at 20, 40, and 75 days after transplanting as affected by various fertilizers and timing of seedling soaking

Treatment	Shoot P uptake (mg pot⁻¹)											
	20 DAT				40 DAT				75 DAT			
	30 min		60 min		30 min		60 min		30 min		60 min	
	Mean	SE	Mean	SE	Mean	SE	Mean	SE	Mean	SE	Mean	SE
Control (+P soil)	1.60 ± 0.5 a		1.05 ± 0.1 ab		2.66 ± 0.7 a		2.51 ± 0.0 a		13.3 ± 2.1 a		12.3 ± 2.6 a	
Control (-P soil)	0.21 ± 0.0 c		0.24 ± 0.0 c		0.43 ± 0.1 b		0.35 ± 0.1 b		1.5 ± 0.5 b		1.7 ± 0.2 b	
1% KH₂PO₄	0.44 ± 0.0 bc		0.39 ± 0.1 bc		0.34 ± 0.0 b		0.63 ± 0.1 b		2.1 ± 0.6 b		3.4 ± 0.7 b	
5% KH₂PO₄	0.80 ± 0.0 bc		0.80 ± 0.1 bc		0.73 ± 0.1 b		0.96 ± 0.1 b		6.3 ± 0.9 b		3.4 ± 0.3 b	
1% NPK	0.25 ± 0.0 c		0.28 ± 0.0 c		0.36 ± 0.1 c		0.40 ± 0.0 b		1.3 ± 0.6 b		0.9 ± 0.2 b	
5% NPK	0.27 ± 0.0 c		-		0.39 ± 0.1 b		-		1.3 ± 0.1 b			

Different letters in the same day after transplanting (DAT) showed significant differences at 0.05% levels by the Tukey-Kramer (n=3)

Table 6. Shoot P uptake of rice plant at 20, 40, and 75 days after transplanting as affected by various fertilizers and timing of seedling soaking

Treatment	Root Dry matter (mg)											
	20 DAT				40 DAT				75 DAT			
	30 min		60 min		30 min		60 min		30 min		60 min	
	Mean	SE	Mean	SE	Mean	SE	Mean	SE	Mean	SE	Mean	SE
Control (+P soil)	171 ± 45 ab		196 ± 22 a		1263 ± 257 ab		1566 ± 141 a		7749 ± 1629 ab		9954 ± 2433 a	
Control (-P soil)	107 ± 26 ab		100 ± 14 ab		216 ± 46 c		205 ± 21 c		1796 ± 644 c		2489 ± 443 c	
1% KH₂PO₄	199 ± 9 a		136 ± 18 ab		219 ± 21 c		472 ± 91 c		1719 ± 439 c		2854 ± 518 bc	
5% KH₂PO₄	166 ± 29 ab		151 ± 21 ab		773 ± 338 bc		935 ± 162 abc		5363 ± 120 abc		3447 ± 602 bc	
1% NPK	114 ± 13 ab		88 ± 11 ab		233 ± 60 c		236 ± 14 c		1379 ± 393 c		1534 ± 270 c	
5% NPK	66 ± 2 b		-		214 ± 18 c		-		1695 ± 308 c			

Different letters in the same day after incubation (DAT) showed significant differences at 0.05% levels by the Tukey-Kramer (n=3)

Table 7. Root dry matter of rice plant at 20, 40, and 75 days after transplanting as affected by various fertilizers and timing of seedling soaking

Treatment	Root P concentration (mg kg⁻¹ DM)											
	20 DAT				40 DAT				75 DAT			
	30 min		60 min		30 min		60 min		30 min		60 min	
	Mean	SE	Mean	SE	Mean	SE	Mean	SE	Mean	SE	Mean	SE
Control (+P soil)	1397 ± 36 a		1038 ± 31 b		730 ± 80 a		521 ± 11 bc		374 ± 16 a		335 ± 32 ab	
Control (-P soil)	560 ± 3 e		739 ± 58 de		513 ± 26 bc		501 ± 25 bc		259 ± 45 abc		227 ± 20 bc	
1% KH₂PO₄	744 ± 53 de		771 ± 27 cde		465 ± 18 c		456 ± 22 c		325 ± 17 ab		255 ± 3 abc	
5% KH₂PO₄	870 ± 65 bcd		998 ± 72 bc		498 ± 19 bc		510 ± 30 bc		252 ± 15 abc		258 ± 8 abc	
1% NPK	743 ± 29 de		692 ± 32 de		557 ± 22 bc		588 ± 36 abc		180 ± 38 c		137 ± 38 c	
5% NPK	742 ± 30 de		-		652 ± 12 ab		-		204 ± 35 bc		-	

Different letters in the same day after transplanting (DAT) showed significant differences at 0.05% levels by the Tukey-Kramer (n=3)

Table 8. Root P concentration of rice plant at 20, 40, and 75 days after transplanting as affected by various fertilizers and timing of seedling soaking

Treatment	Root P uptake (mg pot⁻¹)											
	20 DAT				40 DAT				75 DAT			
	30 min		60 min		30 min		60 min		30 min		60 min	
	Mean	SE	Mean	SE	Mean	SE	Mean	SE	Mean	SE	Mean	SE
Control (+P soil)	0.24 ± 0.1 a		0.20 ± 0.0 ab		0.9 ± 0.1 a		0.8 ± 0.1 ab		2.8 ± 0.5 ab		3.4 ± 0.9 a	
Control (-P soil)	0.06 ± 0.0 c		0.08 ± 0.0 c		0.1 ± 0.0 d		0.1 ± 0.0 d		0.4 ± 0.1 c		0.6 ± 0.1 c	
1% KH₂PO₄	0.15 ± 0.0 abc		0.11 ± 0.0 bc		0.1 ± 0.0 d		0.2 ± 0.0 cd		0.5 ± 0.1 c		0.7 ± 0.1 c	
5% KH₂PO₄	0.14 ± 0.0 abc		0.15 ± 0.0 abc		0.4 ± 0.2 cd		0.5 ± 0.1 bc		1.4 ± 0.1 bc		0.9 ± 0.1 c	
1% NPK	0.08 ± 0.0 c		0.06 ± 0.0 c		0.1 ± 0.0 d		0.1 ± 0.0 cd		0.3 ± 0.1 c		0.2 ± 0.0 c	
5% NPK	0.05 ± 0.0 c		-		0.1 ± 0.0 cd		-		0.3 ± 0.0 c		-	

Different letters in the same day after transplanting (DAT) showed significant differences at 0.05% levels by the Tukey-Kramer (n=3)

Table 9. Root P uptake of rice plant at 20, 40, and 75 days after transplanting as affected by various fertilizers and timing of seedling soaking

Treatment	Height (cm)											
	20 DAT				40 DAT				75 DAT			
	30 min		60 min		30 min		60 min		30 min		60 min	
	Mean	SE	Mean	SE	Mean	SE	Mean	SE	Mean	SE	Mean	SE
Control (+P soil)	53.3 ± 42 a		54.3 ± 12 a		66.7 ± 3 a		65.7 ± 3 a		80.8 ± 1 a		75.0 ± 3 a	
Control (-P soil)	35.2 ± 29 c		35.3 ± 20 c		49.3 ± 4 b		50.0 ± 3 b		62.5 ± 5 b		61.0 ± 2 ab	
1% KH₂PO₄	37.0 ± 25 bc		43.0 ± 12 abc		46.7 ± 1 b		52.0 ± 1 ab		63.5 ± 6 b		66.7 ± 3 ab	
5% KH₂PO₄	45.7 ± 23 abc		48.5 ± 19 abc		52.3 ± 1 ab		54.0 ± 2 ab		69.3 ± 4 ab		66.7 ± 2 ab	
1% NPK	33.2 ± 34 c		36.3 ± 33 bc		44.0 ± 6 b		45.3 ± 2 b		57.8 ± 6 b		56.7 ± 5 b	
5% NPK	34.2 ± 17 c		-		43.0 ± 2 b		-		61.7 ± 4 b		-	

Different letters in the same column showed significant differences at 0.05% levels by the Tukey-Kramer (n=3)

Table 10. Height of rice plant at 20, 40, and 75 days after transplanting as affected by various fertilizers and timing of seedling soaking

Treatment	Leaf age							
	20 DAT				40 DAT			
	30 min		60 min		30 min		60 min	
	Mean	SE	Mean	SE	Mean	SE	Mean	SE
Control (+P soil)	11.1	± 0.4 abcd	11.6	± 0.2 abc	13.4	± 0.3 ab	13	± 0.3 ab
Control (-P soil)	10.1	± 0.2 cd	10.3	± 0.4 bcd	12.7	± 0.2 ab	12	± 0.4 b
1% KH_2PO_4	10.2	± 0.1 cd	10.6	± 0.3 bcd	12.4	± 0.3 b	12	± 0.3 b
5% KH_2PO_4	11.8	± 0.3 ab	12.3	± 0.2 a	14.0	± 0.4 a	14	± 0.2 a
1% NPK	10.7	± 0.3 abcd	10.0	± 0.1 d	12.7	± 0.1 ab	12	± 0.2 b
5% NPK	10.4	± 0.6 bcd	-	-	12.5	± 0.3 b	-	-

Different letters in the same column showed significant differences at 0.05% levels by the Tukey-Kramer (n=3)

Table 11. Leaf age of rice plant at 20 and 40 days after transplanting as affected by various fertilizers and timing of seedling soaking

Treatment	Tiller number							
	20 DAT				40 DAT			
	30 min		60 min		30 min		60 min	
	Mean	SE	Mean	SE	Mean	SE	Mean	SE
Control (+P soil)	4	± 0.3 ab	5	± 0.3 a	5	± 0.9 abc	5	± 1.2 a
Control (-P soil)	1	± 0.0 d	1	± 0.0 d	2	± 0.7 cd	1	± 0.3 d
1% KH_2PO_4	2	± 0.3 cd	3	± 0.3 ab	2	± 0.3 abcd	3	± 0.3 abcd
5% KH_2PO_4	3	± 0.0 bc	4	± 0.3 ab	3	± 0.3 abcd	5	± 1.0 ab
1% NPK	1	± 0.3 d	1	± 0.3 d	2	± 0.3 cd	2	± 0.6 bcd
5% NPK	2	± 0.3 cd	-	-	2	± 0.0 bcd	-	-

Different letters in the same column showed significant differences at 0.05% levels by the Tukey-Kramer (n=3)

Table 12. Tiller number of rice plant at 20 and 40 days after transplanting as affected by various fertilizers and timing of seedling soaking

3.4. Conclusion

Alternative fertilizer utilizing methods have been developed for small subsistence farmers aiming to reduce quantity of fertilizer used to maintain desired level of crop production and re-plenish soil fertility. Although, farmers were aware of soil fertilization but they were not able to access to those fertilizers because of shortage of financial resource. Therefore, alternative meth-ods such as 1) fertilizer microdosing, 2) seed coating, and 3) seedling dipping or soaking have been introduced and the potential of the methods also was provided in this chapter.

Effectiveness of fertilizer microdosing method has been proved on sorghum, millet, and pearl millet production grown on low P fertility, in various soils of severe dry semi-arid and Sahel regions of several countries in Africa. Delayed timing of microdosing still increased crop production and income of small farmers. While, fertilizer seed coating method have been developed to overcome the problem relating to loss of fertilizer during planting after mixing fertilizer with dry crop seed. Fertilizer coating rice seed with use of some adhesive materials resulted to more firm attachment of seed and fertilizer. Early growth and root development of plant was well observed over 40 days after sowing. However, more suitable and affordable adhesive materials and handling procedure should be further invertigated. Dipping or soaking seedling of rice in fertilizer slurry has been traditionally practiced in China. Fertilizers were simply mixed with soil and water to make a paste or slurry and rice seedlings were dipped to this slurry before transplanting. This method reduced fertilizer requirement more than 50%. Moreover, soaking rice seedling by chemical fertilizer such as 5% KH_2PO_4 for 30 min before transplanting could extend the growth of shoot and root up to 75 days on P deficit soils. Alternative fertilizer utilizing methods described above showed relatively high potential for improving the growth of rice seedling, and in turn possibly increased crop productivity in low input agriculture in different soils and climates. These methods were considered as affordable technologies for local subsistence farmer who are not able to access sufficient quantity of fertilizer during cropping season.

Acknowledgement

The authors are grateful to the Ministry of Agriculture, Forestry, and Fisheries (MAFF), Japan for a grant on the Project of 'Improvement of Soil Fertility with Use of Indigenous Resources in Rice Systems of Sub-Sahara Africa' through Japan International Research Center for Agricultural Sciences (JIRCAS).

Author details

Monrawee Fukuda*, Fujio Nagumo*, Satoshi Nakamura* and Satoshi Tobita*

*Address all correspondence to: monrawee@affrc.go.jp

*Address all correspondence to: fnagumo@jircas.affrc.go.jp

*Address all correspondence to: nsatoshi@affrc.go.jp

*Address all correspondence to: bita1mon@jircas.affrc.go.jp

Crop, Livestock, and Environment Division, Japan International Research Center for Agricultural Sciences (JIRCAS), Ohwashi, Tsukuba, Ibaraki, Japan

References

[1] Sanchez PA 2002: Soil fertility and Hunger in Africa. Science, 295, 2019-2020.

[2] Buresh RJ, Smithson PC 1997: Building soil phosphorus capital in Africa. In Replenishing soil fertility in Africa. Buresh RJ, Sanchez PA, and Calhoun F eds. SSSA Special Publication Number 51. Madison, WI, 111-149.

[3] Issaka RN, Buri MM, Tobita S, Nakamura S, and Owusu-Adjei E 2012: Indigenous fertilizing materials to enhance soil productivity in Ghana. In Soil fertility improvement and integrated nutrient management-A global perspective, Walen JK (ed.) InTech. ISBN 978-953-307-945-5, 119-134.

[4] Food and Agriculture Organization of the United Nations 2001: Soil and nutrient management in Sub-Saharan African in support of the soil fertility initiative. Proceedings of the Expert Consultation, Lusaka, Zambia, 6-9 December 1999.

[5] Dobermaan A, Fairhurst TH 2002: Rice straw management. Better crop international, 16, 7-11.

[6] Tobita S, Issaka RN, Buri MM, Fukuda M, and Nakamura S 2012: Indigenous organic resources for improving soil fertility in rice systems in Sub-Saharan Africa. JIRCAS-Research Highlight 2011. Japan International Research Center for Agricultural Sciences (JIRCAS). Access online at http://www.jircas.affrc.go.jp/english/publication/highlights/2011/2011_05.html

[7] Tian G, Brussaard L, Kang BT 1995: An index for assessing the quality of plant residues and evaluating their effects on soil and crop in the (sub-) humid topics. Applied Soil Ecology, 2, 25-32.

[8] JIRCAS 2010: The Study on Improvement on Soil Fertility with Use of Indigenous Resources in Rice Systems of sub-Sahara Africa. Business Report 2009 (February 2010), Tsukuba, Japan.

[9] JIRCAS 2011: The Study on Improvement on Soil Fertility with Use of Indigenous Resources in Rice Systems of sub-Sahara Africa. Business Report 2010 (February 2011), Tsukuba, Japan.

[10] JIRCAS 2012: The Study on Improvement on Soil Fertility with Use of Indigenous Resources in Rice Systems of sub-Sahara Africa. Business Report 2011 (February 2012), Tsukuba, Japan.

[11] Van Kauwenbergh SJ 2010: World phosphate rock reserves and resources. Technical Bulletin IFDC-T-75, IFDC, Alabama, USA, 48 pp.

[12] Simpson Paul G. 1998. Reactive phosphate rocks: their potential role as P fertilizer for Australian Pastures. Technical Bulletin. La Trobe University. Melbourne, Australia.

[13] Chien SH, Prochnow LI, Tu S 2011: Agronomic and environmental aspects of phosphate fertilizers varying in source and solubility: an update review. Nutr. Cycl. Agroecosys., 89, 229-255.

[14] Richardson AE, Lynch JP, Ryan PR, Delhaize E, Smith AF, Smith SE, Harvey PR, Ryan MH, Veneklaas EJ, Lambers H, Oberson A, Culvenor RA, and Simpson RJ 2011: Plant and microbial strategies to improve the phosphorus efficiency of agriculture. Plant Soil, 349, 121-156.

[15] ICRISAT 2009: Fertilizer microdosing-Boosting production in unproductive lands. Access online at http://www.icrisat.org/impacts/impact-stories/icrisat-is-fertilizer-microdosing.pdf

[16] Bagayoko M, Maman N, Pale S, Sirifi S, Taonda SJB, Traore S, Mason SC 2011: Microdose and N and P fertilizer application rates for pearl millet in West Africa. Afr. J. Agric. Res., 6(5), 1141-1150.

[17] Tabo R, Bationa A, Maimouna DK, Hassane O, Koala S 2006: Fertilizer micro-dosing for the prosperity of small-scale farmers in the Sahel-Final report. Global theme on agroecosystems. Report no.23. ICRISAT, 24 pp.

[18] Hayashi K, Abdoulaye T, Gerard B, Bationo 2008: A Evaluation of application timing in fertilizer micro-dosing technology on millet production in Niger, West Africa. Nutr. Cycl. Agroecosyst., 80, 257-265.

[19] Ros C, Bell RW, White PF 2000: Phosphorus seed coating and soaking for improving seedling growth of Oryza sativa (rice) cv. IR66. Seed Sci. & Technol., 28, 391-401.

[20] De Datta SK, Biswas TK, Charoenchamratcheep C 1990: Phosphorus requirement and management for lowland rice. In Phosphorus requirements for sustainable agriculture in Asia and Oceania. Proceeding of a symposium 6-10 March 1989. IRRI, pp. 307-323.

[21] Lu R, Jiang B, Li C 1982: Phosphorus management for submerged rice soils. In Symposia paper II. Proceedings of the 12th international congress of soils science. Indian Society of Soil Science, New Delhi.

[22] Katyal JC 1978: Management of phosphorus in lowland rice. Phosphorus Agric. 73, 21-34.

[23] Ling Y, Lu R 1961: Acta Pedologica Sinica 9, 37.

Permissions

The contributors of this book come from diverse backgrounds, making this book a truly international effort. This book will bring forth new frontiers with its revolutionizing research information and detailed analysis of the nascent developments around the world.

We would like to thank Dr. Roland N. Issaka, for lending his expertise to make the book truly unique. He has played a crucial role in the development of this book. Without his invaluable contribution this book wouldn't have been possible. He has made vital efforts to compile up to date information on the varied aspects of this subject to make this book a valuable addition to the collection of many professionals and students.

This book was conceptualized with the vision of imparting up-to-date information and advanced data in this field. To ensure the same, a matchless editorial board was set up. Every individual on the board went through rigorous rounds of assessment to prove their worth. After which they invested a large part of their time researching and compiling the most relevant data for our readers. Conferences and sessions were held from time to time between the editorial board and the contributing authors to present the data in the most comprehensible form. The editorial team has worked tirelessly to provide valuable and valid information to help people across the globe.

Every chapter published in this book has been scrutinized by our experts. Their significance has been extensively debated. The topics covered herein carry significant findings which will fuel the growth of the discipline. They may even be implemented as practical applications or may be referred to as a beginning point for another development. Chapters in this book were first published by InTech; hereby published with permission under the Creative Commons Attribution License or equivalent.

The editorial board has been involved in producing this book since its inception. They have spent rigorous hours researching and exploring the diverse topics which have resulted in the successful publishing of this book. They have passed on their knowledge of decades through this book. To expedite this challenging task, the publisher supported the team at every step. A small team of assistant editors was also appointed to further simplify the editing procedure and attain best results for the readers.

Our editorial team has been hand-picked from every corner of the world. Their multi-ethnicity adds dynamic inputs to the discussions which result in innovative outcomes. These outcomes are then further discussed with the researchers and contributors who give their valuable feedback and opinion regarding the same. The feedback is then collaborated with the researches and they are edited in a comprehensive manner to aid the understanding of the subject.

Apart from the editorial board, the designing team has also invested a significant amount of their time in understanding the subject and creating the most relevant covers. They scrutinized every image to scout for the most suitable representation of the subject and create an appropriate cover for the book.

The publishing team has been involved in this book since its early stages. They were actively engaged in every process, be it collecting the data, connecting with the contributors or procuring relevant information. The team has been an ardent support to the editorial, designing and production team. Their endless efforts to recruit the best for this project, has resulted in the accomplishment of this book. They are a veteran in the field of academics and their pool of knowledge is as vast as their experience in printing. Their expertise and guidance has proved useful at every step. Their uncompromising quality standards have made this book an exceptional effort. Their encouragement from time to time has been an inspiration for everyone.

The publisher and the editorial board hope that this book will prove to be a valuable piece of knowledge for researchers, students, practitioners and scholars across the globe.

List of Contributors

Eric Owusu Adjei, Roland Nuhu Issaka and Mohammed Moro Buri
CSIR-Soil Research Institute (SRI), Academy Post Office, Kwadaso - Kumasi, Ghana

Vincent Kodjo Avornyo, Joseph A. Awuni and Israel K. Dzomeku
University for Development Studies (UDS) Faculty of Agriculture, Department of Agronomy, Tamale, Ghana

Braulio Valles-de la Mora, Epigmenio Castillo-Gallegos, Jesús Jarillo-Rodríguez and Eliazar Ocaña-Zavaleta
Universidad Nacional Autónoma de México, Facultad de Medicina Veterinaria y Zootecnia, Centro de Enseñanza, Investigación y Extensión en Ganadería Tropical (CEIEGT), México

James M. Kombiok
Savanna Agricultural Research Institute, Tamale, Ghana

Samuel Saaka J. Buah
Savanna Agricultural Research Institute, Wa Station, Wa, Ghana

Jean M. Sogbedji
IFDC, Rue Soloyo, Lome, Togo

Renato de Mello Prado and Gustavo Caione
UNESP (Universidade Estadual Paulista), Jaboticabal, SP, Brazil

S.É. Parent and L.E. Parent
ERSAM, Department of Soils and Agrifood Engineering, Université Laval, Québec (Qc), Canada

D.E. Rozanne
Departamento de Agronomia, Unesp, Universidade Estadual Paulista, Campus de Registro, Registro, São Paulo, Brazil

A. Hernandes and W. Natale
Departamento de Solos e Adubos, Unesp, Universidade Estadual Paulista, Jaboticabal, São Paulo, Brazil

William Natale
Unesp, Universidade Estadual Paulista, Campus Jaboticabal, Via de Acesso Paulo D. Donato Castelane, Jaboticabal, São Paulo, Brazil

Sarita Leonel
UNESP FCA, Department of plant production, Botucatu, SP, Brazil

Luis Lessi dos Reis
UNESP (São Paulo State University), Botucatu-SP, Brazil

Witold Grzebisz, Witold Szczepaniak, Jarosław Potarzycki and Remigiusz Łukowiak
Department of Agricultural Chemistry and Environmental Biogeochemistry, Poznan, University of Life Sciences, Poland

Monrawee Fukuda, Fujio Nagumo, Satoshi Nakamura and Satoshi Tobita
Crop, Livestock, and Environment Division, Japan International Research Center for Agricultural Sciences (JIRCAS), Ohwashi, Tsukuba, Ibaraki, Japan